Basic MATLAB®, Simulink®, and Stateflow®

Basic MATLAB®, Simulink®, and Stateflow®

Richard Colgren
The University of Kansas
Lawrence, Kansas

EDUCATION SERIES
Joseph A. Schetz
Series Editor-in-Chief
Virginia Polytechnic Institute and State University
Blacksburg, Virginia

Published by
American Institute of Aeronautics and Astronautics, Inc.
1801 Alexander Bell Drive, Reston, VA 20191-4344

MATLAB®, Simulink®, Stateflow®, Handle Graphics®, Real-Time Workshop®, and xPC Targetbox® are registered trademarks of The MathWorks, Inc.

American Institute of Aeronautics and Astronautics, Inc., Reston, Virginia

1   2   3   4   5

Library of Congress Cataloging-in-Publication Data
Colgren, Richard D. (Richard Dean).
  Basic MATLAB, Simulink, and Stateflow / Richard Colgren.
      p. cm. -- (Education series)
   ISBN-13: 978-1-56347-838-3
   ISBN-10: 1-56347-838-2 (hardcover : alk. paper)
  1. MATLAB. 2. SIMULINK. 3. Stateflow. 4. Computer simulation—Computer
programs. 5. Engineering mathematics. 6. Aeronautics—Mathematics. I. Title.

   TA345.C598 2007
   620.001'13--dc22

                                                              2006101307

Foreword

We are very happy to present *Basic MATLAB®, Simulink®, and Stateflow®* by Richard Colgren. We are confident that this comprehensive and in-depth treatment of this widely-used material will be very well received by the technical community. The book has fourteen chapters and three appendices in about 500 pages.

This author is extremely well qualified to write this book because of his broad and deep expertise in the area. His command of the material is excellent, and he is able to organize and present it in a very clear manner.

The AIAA Education Series aims to cover a very broad range of topics in the general aerospace field, including basic theory, applications, and design. Information about the complete list of titles can be found on the last page of this volume. The philosophy of the series is to develop textbooks that can be used in a university setting, instructional materials for continuing education and professional development courses, and also books that can serve as the basis for independent study. Suggestions for new topics or authors are always welcome.

Joseph A. Schetz
Editor-in-Chief
AIAA Education Series

Table of Contents

Basic MATLAB®

Software download information can be found at the end of
the book on the Supporting Materials page.

Preface

This book is based on materials developed during more than 22 years of teaching MATLAB®, Simulink®, and Stateflow® in a variety of formats to a diverse range of audiences. Most of these courses required little to no background in any of these tools from these students. The book can be used for self-instruction on all three of these topics. All of these tools are relatively easy to use once the basics are understood. The hands-on approach taken in this book is designed to provide the user with just such a background. This book is in no way meant to be comprehensive in its coverage of these three tool sets. A comprehensive book on MATLAB was possible 20 years ago. However, with the vast number of toolboxes and model libraries available today, a comprehensive coverage of these three subjects would require a bookshelf, not a single book.

When used as a classroom text, this book is formatted to support a MATLAB/Simulink/Stateflow course designed to take a total of approximately 40 hours, including in-class exercises. However, the course is designed to be modular and thus flexible for use in a variety of teaching and time formats. As a class, this course is best offered within a computer laboratory environment, with the students working in real time on examples along with the instructor.

Note that all the materials covered within this book were generated using Version 7.3 of MATLAB (Release 2006b) and all associated toolbox versions. The Math-Works is now supporting a twice-yearly release schedule, with each Service Pack providing minor upgrades and some new features. These added improvements are designed to have little effect on the vast majority of capabilities offered by this large family of analysis tools and thus are relatively seamless to the user.

The format recommended for this course as offered within a computer laboratory environment is as follows.

1) A lecture is given on a MATLAB, Simulink, or Stateflow topic from the appropriate chapter in this book. The students work through the materials on their computers while the instructor similarly works through the lecture materials.

2) The students work through an exercise given at the end of the chapter after the lecture.

3) An appropriate break is given at the end of the exercise. Students completing the exercise early may work on other side topics.

4) Afterward the instructor assigns exercises or homework for the students to complete.

Richard Colgren
January 2007

Acknowledgments

In my more than 25 years of working with MATLAB®, Simulink®, and State-flow® and all of the associated tools, it has been my pleasure to have had the assistance of, and to have received advice and wisdom from, numerous people. It is my great pleasure to acknowledge all of them, and my greatest fear is that I will miss giving the appropriate credit to one or more of these deserving people. To any of you I have missed, I first state my appreciation for your support and then ask your forgiveness for missing you.

My first acknowledgment must go to my wife, Nell, for all of her support and her understanding during the time I have spent teaching, traveling, and writing. The time she has given up to support these efforts is greatly appreciated, and her love and patience are acknowledged. I must also express my gratitude to my parents, who have done so much to support my education and my educational activities over the years.

My first exposure to MATLAB was in the 1980–81 school year at the University of Washington. The excellent facilities at the university, along with the excellent engineering faculty, provided me with an early opportunity to work with these tools and had a profound effect on my career.

At Northrop from 1982 through 1984, I was given detailed insight into the workings of MATLAB. The person who most greatly contributed to this insight was Dave Lowry. At Northrop I was first exposed to and received training on a very early release of Matrix-X, a program that was then very closely related to the university version of MATLAB. My departure from Northrop and arrival at Lockheed happened to be almost coincidental with the founding of The MathWorks.

My first task at Lockheed in mid-April 1984 was to participate in a review of Ctrl-C, another package very closely related to MATLAB. This review first brought me into contact with Jack Little, who is currently president of The Math-Works. This work then lead me to order (for $75) two nine-track tapes with MATLAB and computer-specific interface programs from Cleve Moler at the University of New Mexico. Moler developed the original FORTRAN version of MATLAB and is currently the chief scientist of The MathWorks.

Two of my managers at the Lockheed "Skunk Works," Bob "Lash" Loschke and Bob Rooney (both now retired), provided me with the support and encouragement to implement MATLAB and associated training programs at Lockheed. My first task was to implement MATLAB on the "batch" mainframe computers of the time and to provide an apparently "interactive" interface including color plotting capabilities. With the support of the small IS group at the Lockheed Skunk Works, I wrote dynamic JCL code and assembly language interfaces to accomplish this task.

When MATLAB became commercially available on IBM PCs and Apple Macintosh computers, these computers were brought into Lockheed along with this software. With the aid of Hank Donald, then an engineer at Lockheed and now at Ford (where he has continued to work on MATLAB, Simulink, and Stateflow standards), we brought UNIX workstations and the next generation of MATLAB and Matrix-X based tools into the Skunk Works. These workstations were also used to introduce us to early graphical modeling tools including Grumman's Protoblock, ISI's SystemBuild, X-ANALOG's NL-SIM, and ADI's BEACON.

The introduction of MATLAB and additional software tools by The MathWorks led to the need for additional training modules and instructors. Hank Donald helped generate the initial training materials for MEX-files. His desire for a manual Simulink switch and other features within Simulink really motivated The MathWorks' developers to improve the early versions of Simulink and make it the excellent tool you have today. He also participated in many of the early training courses within Lockheed. Shah Torgenson wrote some of the earliest training materials on cells and structures and helped tremendously with several of the training classes we offered at Lockheed. Bob Radford and Bill Wood were also very helpful in their support and participation in several MATLAB training classes at Lockheed. The human resources department at Lockheed and later Lockheed Martin helped support some of this work. Jim Buffington and his engineers, including Mike Niestroy in Fort Worth, along with the F-22 program, including Dave Seto, also provided great help in expanding this training throughout Lockheed Martin.

I appreciated The MathWorks providing a copy of Simulab for early evaluation work. This software then developed into the Simulink modeling package. Similarly, The MathWorks provided me with the first version of Stateflow software, which was then not integrated into Simulink. All following versions of Stateflow software were integrated within the Stateflow environment.

Many others have provided me with assistance at The MathWorks. Again, I apologize for those that I have missed. Russell Scarlata, who had the Lockheed account at The MathWorks, provided me with years of excellent help and support. John Binder also provided several years of excellent support and encouraged me to develop this book. Dick Gram, previously at Grumman and then at The MathWorks, provided many great flight controls modeling ideas. Courtney Esposito, of the MATLAB Book Program, has provided great software and publishing support for my previous book, *Applications of Robust Control to Nonlinear Systems*, as well as this book.

I would like to thank the University of Kansas Aerospace Engineering Short Course Program for aiding me in offering MATLAB and Simulink as both an on-site class and as a course offered to the general public. Finally, I would like to thank the staff of the Department of Aerospace Engineering at the University of Kansas for their support in my writing and completing this book.

Basic MATLAB®

1
Introduction to MATLAB®

1.1 Introduction and Objectives

MATLAB® is a high-performance language for technical computing. It integrates computation, visualization, and programming in an easy-to-use environment where problems and solutions are expressed in familiar mathematical notation. This chapter introduces some of the basic matrix computational tools and graphical user interfaces (GUIs) that are available in MATLAB through the main **Command Window**.

Upon completion of this chapter, the reader will be able to 1) identify some basic computational tools and commands in MATLAB; 2) identify the various components of the MATLAB GUI; 3) input commands into the **Command Window**; 4) obtain help information using on-line help utilities, local contacts, users groups, and The MathWorks Web site and help lines.

On Windows platforms, to start MATLAB, double-click the MATLAB shortcut icon on your Windows desktop (see Fig. 1.1).

Fig. 1.1

On UNIX platforms, to start MATLAB, type **matlab** at the operating system prompt.

After you have started the MATLAB program, the standard main MATLAB interface window appears as shown in Fig. 1.2.

You can change the way your desktop looks by opening, closing, moving, and resizing the tools on it. Use the **View** menu to open or close the tools. You can also move tools outside the desktop or move them back onto the desktop. All the desktop tools provide common features such as context menus and keyboard shortcuts. You can specify certain characteristics for the desktop tools by selecting **Preferences** from the **File** menu. For example, you can specify the font characteristics for **Command Window** text. For more information on this or any topic, click the **Help** button in the **Preferences** dialog box.

3

Fig. 1.2

In the discussion to follow, the given commands will be entered using the **Command Window**. Statements you enter into the **Command Window** are logged into the **Command History**. In the **Command History** window, you can view previously run statements and copy and execute selected statements. You can also use the up and down arrows on your keyboard to place previous commands directly into the **Command Window** for execution.

You can run external programs from the MATLAB **Command Window**. The exclamation point ! indicates that the rest of the input line is a command to the operating system. This is useful for invoking utilities or running other programs without quitting MATLAB. The MATLAB **Start** button ⚙Start provides easy access to tools, demos, and documentation. Just click on the button to see the options.

MATLAB file operations use the current directory and the search path as reference points. Any file you want to run must either be in the current directory or on the search path. A quick way to view or change the current directory is by using the **Current Directory** field in the desktop toolbar.

The MATLAB workspace consists of arrays or matrices generated during your MATLAB session and stored in memory. You add variables to the workspace by using functions, running M-files, and loading saved workspaces.

1.2   Entry

To enter a matrix, spaces or commas are put between the elements. Semicolons or returns are used to separate the rows. Note that semicolons at the end of a command suppress the echo print. Brackets are placed around the matrix data. For example, to enter a 3-by-3 matrix A, type

$\gg$ A = [1 2 3;4 5 6;7 8 0]

which results in

A =
```
   1   2   3
   4   5   6
   7   8   0
```

Typing

$\gg$ A = [1   2   3   or   A = [1, 2, 3
         4   5   6   4, 5, 6
         7   8   0]   7, 8, 0]

produces the same results. Remember that a scalar is just a 1-by-1 matrix. Typing

$\gg$ **e = 1**

produces the result

e =
```
   1
```

MATLAB does not require that numbers be declared as real, integer, et cetera, nor does it require that matrices be dimensioned. The software has very good algorithms for deducing variable types. This will be covered in more detail later.

1.3 Transpose

Next we will run through some basic matrix operations. They are done similar to the way you might write them on paper. For example, the matrix A can be transposed with the command

$\gg$ **B = A'**

which results in the new matrix B

B =
```
   1   4   7
   2   5   8
   3   6   0
```

Remember that transposing a matrix exchanges its rows and columns. This is a useful operation in matrix mathematics.

1.4 Addition and Subtraction

Matrix addition and subtraction are done element by element. Note that matrices must be of the same dimension for this to be valid (unless subtracting a scalar value from a matrix). Many program errors using matrix addition and subtraction fail because of improper dimensioning. MATLAB has tools for checking matrix dimensions, etc. More later.

Adding our previously generated matrices

≫ **C = A + B**

gives

```
C =
     2    6   10
     6   10   14
    10   14    0
```

Note that $A(1,1) = 1$, $B(1,1) = 1$, and so $C(1,1) = A(1,1) + B(1,1) = 1 + 1 = 2$, etc.

Subtracting our previously generated matrices

≫ **C = A − B**

gives

```
C =
     0   −2   −4
     2    0   −2
     4    2    0
```

Note that $A(1,1) = 1$, $B(1,1) = 1$, and so $C(1,1) = A(1,1) − B(1,1) = 1 − 1 = 0$, etc.

What does **C = A−e** give? It subtracts the scalar **e** from every element in **A** and saves it as the matrix **C**:

≫ **C = A − e**

gives

```
C =
     0    1    2
     3    4    5
     6    7   −1
```

Note that reusing the same name **C** has caused the old element data to be overwritten. Matrices can be saved using the **save** command. The command **save temp C** would save the current **C** in the file temp.mat. To read in the resulting MAT-file, select **Import Data** from the **File** menu, or use the **load temp** command. **Save Workspace As** from the **File** menu saves everything in the workspace to the specified MAT-file, which has a **.mat** extension. If an expression is evaluated and no variable is assigned to the result, it is saved as **ans**. To see what values we have generated, type

   ≫ **who**

which results in
   Your variables are

   A   B   C
   e

The command **whos** gives a table of information including matrix size, number of elements, number of bytes used, density of the matrix, and whether it is complex. This is the same information as that given in the **Workspace** window. The command **clear** removes all these data from memory.

1.5 Multiplication

Both matrix and array multiplications are supported in MATLAB. Operations are defaulted to matrix operations unless denoted by a period **.** as discussed in the following two sections.

1.5.1 Matrix Multiplication

Matrix multiplication is indicated with the use of an asterisk *****.

   ≫ **C = A * B**

producing the new matrix C

   C =
     14   32   23
     32   77   68
     23   68   113

Note that $C(1,1) = A(1,1)*B(1,1) + A(1,2)*B(2,1) + A(1,3)*B(3,1) = 1*1 + 2$
$2 + 3*3 = 2 + 4 + 9 = 14$, etc. This means that the order of multiplication is important (i.e., **A*B** and **B*A** gives different results).

1.5.2 Array Multiplication

Array multiplication is the multiplication of every element in the array by a scalar value and is indicated with a period before the multiplication asterisk as follows:

   ≫ **D = A.* 2**

producing the new matrix **D**

D =
```
 2    4   6
 8   10  12
14   16   0
```

1.6 Division

As with multiplication, both matrix and array division are supported in MATLAB. Operations are defaulted to matrix operations unless denoted by a period as in the operation ./ as discussed in the following two sections. Note that both right and left divisions are supported and that they usually do not result in the same answer.

1.6.1 Matrix Division

Matrix division is indicated with

≫ X = A/B

the solution to X*B = A (right division)

X =
```
-0.3333   0.6667   -0.0000
-3.3333   3.6667   -0.0000
-5.3333   4.6667    1.0000
```

or

≫ X = A \ B

the solution to A*X = B (left division)

X =
```
-0.3333   -3.3333   -5.3333
 0.6667    3.6667    4.6667
-0.0000   -0.0000    1.0000
```

1.6.2 Array Division

Array division is the division of every element in the array by a scalar value and is indicated with a period before the multiplication asterisk as follows:

≫ E = A./2

divides every element in A by 2 (right division)

E =

0.5000	1.0000	1.5000
2.0000	2.5000	3.0000
3.5000	4.0000	0

or

>> **F = A.\2**

divides 2 by each element in A (left division), or $fij = 2/aij$. Note that $2/0 = $ Inf, giving a divide by zero warning as follows:
    Warning: Divide by zero.

F =

2.0000	1.0000	0.6667
0.5000	0.4000	0.3333
0.2857	0.2500	Inf

Double-click any variable in the **Workspace** browser to see it in the **Array Editor**. You can use the **Array Editor** to view and edit a visual representation of one- or two-dimensional numeric arrays, strings, and cell arrays of strings that are in the workspace.

1.7 Formats

MATLAB has changed the representation from an integer display to a real display using "**short** format." **Long** formats (double precision), scientific notation (both **short** and **long**), **bank** (two decimal places), **hex**, and sign (+, −, or blank for a 0 element) are available. The X(1,1) element in the previous matrix can be displayed in each of these formats as follows:

>> **format short**

  −0.3333

>> **format short e**

  −3.3333e-01

>> **format long**

  −0.33333333333333

>> **format long e**

  −3.333333333333333e-01

>> **format bank**

−0.33

>> **format hex**

bfd5555555555555

>> **format +**

−

1.8 Matrix Functions

MATLAB provides four functions that generate basic matrices: 1) **zeros** (matrix of all zeros), 2) **ones** (matrix of all ones), 3) **eye** (matrix of all zeros except for ones along the diagonal), 4) **rand** (matrix of uniformly distributed random elements), and 5) **randn** (matrix of normally distributed random elements). A family of functions are available to calculate common matrix properties and factorizations.

1.8.1 Determinant

As used in matrix inversion

>> **det(A)**

ans =
    27

1.8.2 Rank

Number of independent rows/columns

>> **rank(A)**

ans =
    3

1.8.3 Condition Number

Measures sensitivity to data errors, ratio of largest to smallest singular value

>> **cond(A)**

ans =
    35.1059

1.8.4 Matrix Inverse

Solves A*inv(A) = [I], which is equivalent to 1/e for the scalar case. MATLAB uses this in solving the matrix division problem:

>> **inv(A)**

ans =
        −1.7778   0.8889   −0.1111
         1.5556   −0.7778   0.2222
        −0.1111   0.2222   −0.1111

1.8.5 Eigenvalues

Gives the nontrivial solutions to the problem $Ax = \lambda x$. The n values of λ are the eigenvalues, and the corresponding values of x are the right eigenvectors:

≫ **eig(A)**

ans =
        12.1229
        −0.3884
        −5.7345

We can obtain eigenvectors as well as eigenvalues if we use two arguments on the left-hand side:

≫ **[v,d] = eig(A)**

v =
        0.7471   −0.2998   −0.2763
       −0.6582   −0.7075   −0.3884
        0.0931   −0.6400   0.8791

d =
       −0.3884   0   0
          0   12.1229   0
          0   0   −5.7345

The left eigenvalues (satisfying $WA = DW$) can be computed by using the two statements

≫ **[W,D] = eig(A'), W = W'**

The eigenvectors in **D** computed from **W** and **W'** are the same, although they may occur in different orders.

The generalized eigenvalue problem is the solution to $Ax = \lambda Bx$, where A and B are both n-by-n matrices. The values of λ that satisfy the equation are the generalized eigenvalues, and the corresponding values of x are the generalized right eigenvectors. If B is nonsingular, the problem could be solved by reducing it to a standard eigenvalue problem $inv(B)*A*x = \lambda*x$. The command

≫ **[V,D] = eig(A,B)**

produces the diagonal matrix D of generalized eigenvalues and the full matrix V whose columns are the corresponding eigenvectors so that $A*V = B*V*D$.

The eigenvectors are scaled so that the norm of each is 1.

Note that for help about any command in MATLAB you can type the following,

≫ **help [command]**

to display information about that command; e.g.,

≫ **help eig**

MATLAB provides many more advanced mathematical functions, including Bessel and gamma functions. Most of these functions accept complex arguments.

For a list of the elementary mathematical functions, type **help elfun**. For a list of more advanced mathematical and matrix functions, type **help specfun** and **help elmat**.

Some of the functions, like **sqrt** and **sin**, are built into MATLAB. They are part of the MATLAB core, and so they are very efficient, but the computational details are not readily accessible. Other functions, like **gamma** and **sinh**, are implemented as M-files. You can see the code and even modify the code if you want.

1.8.6 Singular Value Decomposition

Uses the QR method to produce the diagonal matrix S and unitary matrices U and V satisfying $A = U*S*V'$. Operator by itself gives the vector of diagonal elements of S:

≫ **svd(A)**

ans =
    13.2015
    5.4388
    0.3760

The full command **[U,S,V] = svd(A)** gives the full matrices U, S, and V.

1.8.7 Matrix Exponential

Generates the exponent of a matrix:

≫ **expm(A)**

ans =
    1.0e + 04*
    3.1591   3.9741   2.7487
    7.4540   9.3775   6.4858
    6.7431   8.4830   5.8672

Element-wise exponentials are calculated using the command **exp(A)**:

≫ **exp(A)**

    1.0e + 03*
    0.0027   0.0074   0.0201
    0.0546   0.1484   0.4034
    1.0966   2.9810   0.0010

1.8.8 Characteristic Polynomial

Gives the $n + 1$ element row vector whose elements are the coefficients of the characteristic polynomial **det(sI − A)**. The roots of this polynomial are the eigenvalues of the matrix A:

≫ **p = poly(A)**
p =
    1.0000   −6.0000   −72.0000   −27.0000

The elements of this vector represent the polynomial coefficients in descending powers.

The roots of this polynomial are, of course, the eigenvalues of A:

```
>> roots(p)
ans =
        12.1229
        -5.7345
        -0.3884

>> eig(A)
ans =
        12.1229
        -0.3884
        -5.7345
```

For vectors, roots and poly are inverse functions of each other, up to ordering, scaling, and roundoff error.

1.9 Colon Operator

The colon : is one of the most important MATLAB operators. It occurs in several forms. The expression **1:10** is a row vector containing integers from 1 to 10:

```
1   2   3   4   5   6   7   8   9   10
```

To obtain nonunit spacing, specify an increment. For example, **100: −7:50** is

```
100   93   86   79   72   65   58   51
```

Subscript expressions involving colons refer to portions of a matrix. For example, **A(1:k,j)** is the first k elements of the jth column of **A**.

1.10 Useful Interface GUIs

The basic MATLAB demonstrations can be started by selecting the MATLAB **Start** button [▲Start]. You can also type **demo** in the Command Window to open the **Help** browser to the **Demos** tab or go directly to the demos for a specific product or category. For example, **demo matlab graphics** lists the demos for MATLAB Graphics. MATLAB's help and on-line documentation can always be started by selecting **MATLAB Help** from the **Help** pull-down menu. The **Help** interface appears as shown in Fig. 1.3.

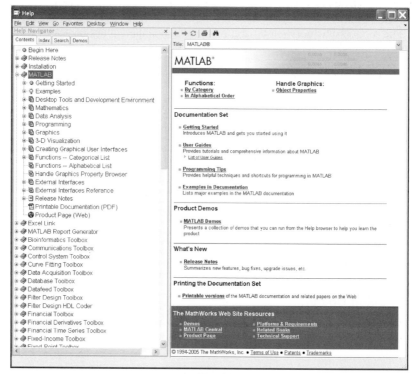

Fig. 1.3

Help is also available at the command line level. Help "topic" gives you help for the topic of interest. Typing

>> **help help**

provides the following information:

HELP On-line help, display text at command line.

HELP, by itself, lists all primary help topics. Each primary topic corresponds to a directory name on the MATLABPATH.

HELP TOPIC gives help on the specified topic. The topic can be a command name, a directory name, or a MATLABPATH relative partial path name (see HELP PARTIALPATH). If it is a command name, HELP displays information on that command. If it is a directory name, HELP displays the Table-Of-Contents for the specified directory. For example, **help general** and **help matlab/general** both list the Table-Of-Contents for the directory toolbox/matlab/general.

HELP FUN displays the help for the function FUN.

T = help ('topic')

returns the help text in an /n separated string.

LOOKFOR XYZ looks for the string XYZ in the first comment line of the HELP text in all M-files found on the MATLABPATH. For all files in which a match occurs, LOOKFOR displays the matching lines.

MORE ON causes HELP to pause between screenfuls if the help text runs to several screens.

In the on-line help, keywords are capitalized to make them stand out. Always type commands in lowercase because all command and function names are actually in lowercase.

For tips on creating help for your M-files, type **help.m**.

See also LOOKFOR, WHAT, WHICH, DIR, MORE.

Overloaded methods
   help cvtest/help.m
   help cvdata/help.m

Help by itself gives a list of all the available help topics, as follows:

  ≫ **help**

HELP topics

help
HELP topics

matlab\general	– General purpose commands.
matlab\ops	– Operators and special characters.
matlab\lang	– Programming language constructs.
matlab\elmat	– Elementary matrices and matrix manipulation.
matlab\elfun	– Elementary math functions.
matlab\specfun	– Specialized math functions.
matlab\matfun	– Matrix functions-numerical linear algebra.
matlab\datafun	– Data analysis and Fourier transforms.
matlab\polyfun	– Interpolation and polynomials.
matlab\funfun	– Function functions and ODE solvers.
matlab\sparfun	– Sparse matrices.
matlab\scribe	– Annotation and Plot Editing.
matlab\graph2d	– Two dimensional graphs.
matlab\graph3d	– Three dimensional graphs.
matlab\specgraph	– Specialized graphs.
matlab\graphics	– Handle Graphics.
matlab\uitools	– Graphical user interface tools.
matlab\strfun	– Character strings.
matlab\imagesci	– Image and scientific data input/output.
matlab\iofun	– File input and output.
matlab\audiovideo	– Audio and Video support.
matlab\timefun	– Time and dates.
matlab\datatypes	– Data types and structures.
matlab\verctrl	– Version control.
matlab\codetools	– Commands for creating and debugging code.
matlab\helptools	– Help commands.
matlab\winfun	– Windows Operating System Interface Files (COM/DDE)
matlab\demos	– Examples and demonstrations.

toolbox\local	– Preferences.
simulink\simulink	– Simulink
simulink\blocks	– Simulink block library.
simulink\components	– Simulink components.
simulink\fixedandfloat	– Simulink Fixed Point utilities.
fixedandfloat\fxpdemos	– Fixed-Point Blockset Demos
fixedandfloat\obsolete	– (No table of contents file)
simulink\simdemos	– Simulink 4 demonstrations and samples.
simdemos\aerospace	– Simulink: Aerospace model demonstrations and samples.
simdemos\automotive	– Simulink: Automotive model demonstrations and samples.
simdemos\simfeatures	– Simulink: Feature demonstrations and samples.
simfeatures\mdlref	– (No table of contents file)
simdemos\simgeneral	– Simulink: General model demonstrations and samples.
simdemos\simnew	– Simulink: New features model demonstrations and samples.
simulink\dee	– Differential Equation Editor
shared\dastudio	– (No table of contents file)
stateflow\stateflow	– Stateflow
rtw\rtw	– Real-Time Workshop
shared\hds	– (No table of contents file)
shared\timeseries	– Shared Time Series Toolbox Library
stateflow\sfdemos	– Stateflow demonstrations and samples.
stateflow\coder	– Stateflow Coder
rtw\rtwdemos	– Real-Time Workshop Demos
rtwdemos\rsimdemos	– (No table of contents file)
asap2\asap2	– (No table of contents file)
asap2\user	– (No table of contents file)
rtwin\rtwin	– Real-Time Windows Target
simulink\accelerator	– Simulink Accelerator
rtw\accel	– (No table of contents file)
aeroblks\aeroblks	– Aerospace Blockset
aeroblks\aerodemos	– Aerospace Blockset demonstrations and examples.
aerodemos\texture	– (No table of contents file)
bioinfo\bioinfo	– Bioinformatics Toolbox
bioinfo\microarray	– Bioinformatics Toolbox—Microarray support functions.

bioinfo\proteins	– Bioinformatics Toolbox—Protein analysis tools.
bioinfo\biomatrices	– Bioinformatics Toolbox—Sequence similarity scoring matrices.
bioinfo\biodemos	– Bioinformatics Toolbox—Tutorials, demos and examples.
can\blocks	– (No table of contents file)
configuration\resource	– (No table of contents file)
common\tgtcommon	– (No table of contents file)
c166\c166	– Embedded Target for Infineon C166 Microcontrollers
c166\blocks	– (No table of contents file)
c166\c166demos	– (No table of contents file)
ccslink\ccslink	– Link for Code Composer Studio
ccslink\ccsblks	– (No table of contents file)
ccslink\ccsdemos	– Link for Code Composer Studio® Demos
cdma\cdma	– CDMA Reference Blockset
cdma\cdmamasks	– CDMA Reference Blockset mask helper functions.
cdma\cdmamex	– CDMA Reference Blockset S-Functions.
cdma\cdmademos	– CDMA Reference Blockset demonstrations and examples.
comm\comm	– Communications Toolbox
comm\commdemos	– Communications Toolbox Demonstrations.
commdemos\commdocdemos	– Communications Toolbox Documentation Examples.
comm\commobsolete	– Archived MATLAB Files from Communications Toolbox Version 1.5.
commblks\commblks	– Communications Blockset
commblks\commmasks	– Communications Blockset mask helper functions.
commblks\commmex	– Communications Blockset S-Functions.
commblks\commblksdemos	– Communications Blockset Demos.
commblksobsolete\v2p5	– (No table of contents file)
commblksobsolete\v2	– (No table of contents file)
commblksobsolete\v1p5	– Archived Simulink Files from Communications Toolbox Version 1.5.
control\control	– Control System Toolbox
control\ctrlguis	– Control System Toolbox—GUI support functions.
control\ctrlobsolete	– Control System Toolbox—obsolete commands.
control\ctrlutil	– (No table of contents file)

control\ctrldemos	– Control System Toolbox—Demos.
shared\controllib	– Control Library
curvefit\curvefit	– Curve Fitting Toolbox
curvefit\cftoolgui	– (No table of contents file)
shared\optimlib	– Optimization Library
daq\daq	– Data Acquisition Toolbox
daq\daqguis	– Data Acquisition Toolbox—Data Acquisition Soft Instruments.
daq\daqdemos	– Data Acquisition Toolbox—Data Acquisition Demos.
database\database	– Database Toolbox
database\dbdemos	– Database Toolbox Demonstration Functions.
database\vqb	– Visual Query Builder functions.
datafeed\datafeed	– Datafeed Toolbox
datafeed\dfgui	– Datafeed Toolbox Graphical User Interface
drive\drive	– SimDriveline
drive\drivedemos	– (No table of contents file)
dspblks\dspblks	– Signal Processing Blockset
dspblks\dspmasks	– Signal Processing Blockset mask helper functions.
dspblks\dspmex	– DSP Blockset S-Function MEX-files.
dspblks\dspdemos	– Signal Processing Blockset demonstrations and examples.
targets\ecoder	– Real-Time Workshop Embedded Coder
ecoder\ecoderdemos	– (No table of contents file)
targets\mpt	– (No table of contents file)
mpt\mpt	– Module Packaging Tool
mpt\user_specific	– (No table of contents file)
toolbox\exlink	– Excel Link
symbolic\extended	– Extended Symbolic Math
filterdesign\filterdesign	– Filter Design Toolbox
filterdesign\quantization	– (No table of contents file)
filterdesign\filtdesdemos	– Filter Design Toolbox Demonstrations.
finance\finance	– Financial Toolbox
finance\calendar	– Financial Toolbox calendar functions.
finance\findemos	– Financial Toolbox demonstration functions.
finance\finsupport	– (No table of contents file)
finderiv\finderiv	– Financial Derivatives Toolbox
finfixed\finfixed	– Fixed-Income Toolbox
fixedpoint\fixedpoint	– Fixed-Point Toolbox
fixedpoint\fidemos	– (No table of contents file)
fixedpoint\fimex	– (No table of contents file)
toolbox\fixpoint	– Simulink Fixed Point

ftseries\ftseries	– Financial Time Series Toolbox
ftseries\ftsdemos	– (No table of contents file)
ftseries\ftsdata	– (No table of contents file)
ftseries\ftstutorials	– (No table of contents file)
fuzzy\fuzzy	– Fuzzy Logic Toolbox
fuzzy\fuzdemos	– Fuzzy Logic Toolbox Demos.
toolbox\gads	– (No table of contents file)
gads\gads	– Genetic Algorithm Direct Search Toolbox
gads\gadsdemos	– Genetic Algorithm Direct Search Toolbox
garch\garch	– GARCH Toolbox
garch\garchdemos	– (No table of contents file)
toolbox\gauges	– Gauges Blockset
hc12\hc12	– Embedded Target for Motorola HC12
hc12\blocks	– (No table of contents file)
hc12\codewarrior	– (No table of contents file)
hc12\hc12demos	– (No table of contents file)
hdlfilter\hdlfilter	– Filter Design HDL Coder
hdlfilter\hdlfiltdemos	– (No table of contents file)
shared\hdlshared	– HDL Library
ident\ident	– System Identification Toolbox
ident\idobsolete	– (No table of contents file)
ident\idguis	– (No table of contents file)
ident\idutils	– (No table of contents file)
ident\iddemos	– (No table of contents file)
ident\idhelp	– (No table of contents file)
images\images	– Image Processing Toolbox
images\imuitools	– Image Processing Toolbox—imuitools
images\imdemos	– Image Processing Toolbox—demos and sample images
images\iptutils	– Image Processing Toolbox utilities
imaq\imaq	– Image Acquisition Toolbox
imaq\imaqdemos	– Image Acquisition Toolbox.
imaqblks\imaqblks	– (No table of contents file)
imaqblks\imaqmasks	– (No table of contents file)
imaqblks\imaqmex	– (No table of contents file)
instrument\instrument	– Instrument Control Toolbox
instrument\instrumentdemos	– (No table of contents file)
instrumentblks\instrumentblks	– (No table of contents file)
instrumentblks\instrumentmex	– (No table of contents file)
map\map	– Mapping Toolbox
map\mapdemos	– Mapping Toolbox Demos and Data Sets.
map\mapdisp	– Mapping Toolbox Map Definition and Display.

map\mapformats	– Mapping Toolbox File Formats.
map\mapproj	– Mapping Toolbox Projections.
mbc\mbc	– Model-Based Calibration Toolbox
mbc\mbcdata	– Model-Based Calibration Toolbox.
mbc\mbcdesign	– Model-Based Calibration Toolbox.
mbc\mbcexpr	– Model-Based Calibration Toolbox.
mbc\mbcguitools	– Model-Based Calibration Toolbox.
mbc\mbclayouts	– (No table of contents file)
mbc\mbcmodels	– Model-Based Calibration Toolbox.
mbc\mbcsimulink	– Model-Based Calibration Toolbox.
mbc\mbctools	– Model-Based Calibration Toolbox.
mbc\mbcview	– Model-Based Calibration Toolbox.
mech\mech	– SimMechanics
mech\mechdemos	– SimMechanics Demos.
pmimport\pmimport	– (No table of contents file)
modelsim\modelsim	– Link for ModelSim
modelsim\modelsimdemos	– (No table of contents file)
mpc\mpc	– Model Predictive Control Toolbox
mpc\mpcdemos	– (No table of contents file)
mpc\mpcguis	– (No table of contents file)
mpc\mpcobsolete	– Contents of the previous (obsolete) version of the MPC Toolbox
mpc\mpcutils	– (No table of contents file)
shared\slcontrollib	– Simulink Control Design Library
targets\mpc555dk	– (No table of contents file)
common\configuration	– (No table of contents file)
mpc555dk\mpc555demos	– (No table of contents file)
mpc555dk\mpc555dk	– Embedded Target for Motorola MPC555
mpc555dk\pil	– (No table of contents file)
blockset\mfiles	– (No table of contents file)
rt\blockset	– (No table of contents file)
nnet\nnet	– Neural Network Toolbox
nnet\nnutils	– (No table of contents file)
nnet\nncontrol	– Neural Network Toolbox Control System Functions.
nnet\nndemos	– Neural Network Demonstrations.
nnet\nnobsolete	– (No table of contents file)
opc\opc	– OPC Toolbox
opc\opcgui	– (No table of contents file)
opc\opcdemos	– (No table of contents file)
toolbox\optim	– Optimization Toolbox
osek\osek	– Embedded Target for OSEK VDX
osek\osekdemos	– (No table of contents file)

osek\blocks	– (No table of contents file)
osek\osekworks	– (No table of contents file)
osek\proosek	– (No table of contents file)
toolbox\pde	– Partial Differential Equation Toolbox
powersys\powersys	– SimPowerSystems
powersys\powerdemo	– SimPowerSystems Demos
drives\drives	– (No table of contents file)
drives\drivesdemo	– (No table of contents file)
facts\facts	– (No table of contents file)
facts\factsdemo	– (No table of contents file)
DR\DR	– (No table of contents file)
DR\DRdemo	– (No table of contents file)
simulink\reqmgt	– (No table of contents file)
reqmgt\rmidemos	– (No table of contents file)
rf\rf	– RF Toolbox
rf\rfdemos	– RF Toolbox Demos.
rf\rftool	– RF Tool (GUI)
rfblks\rfblks	– RF Blockset
rfblks\rfblksmasks	– RF Blockset mask helper functions.
rfblks\rfblksmex	– RF Blockset S-Functions.
rfblks\rfblksdemos	– RF Blockset Demos.
robust\robust	– Robust Control Toolbox
robust\rctutil	– (No table of contents file)
robust\rctdemos	– (No table of contents file)
rctobsolete\lmi	– Robust Control Toolbox-LMI Solvers.
mutools\commands	– (No table of contents file)
mutools\subs	– (No table of contents file)
rptgen\rptgen	– MATLAB Report Generator
rptgen\rptgendemos	– (No table of contents file)
rptgenext\rptgenext	– Simulink Report Generator
rptgenext\rptgenextdemos	– (No table of contents file)
signal\signal	– Signal Processing Toolbox
signal\sigtools	– (No table of contents file)
signal\sptoolgui	– (No table of contents file)
signal\sigdemos	– Signal Processing Toolbox Demonstrations.
slcontrol\slcontrol	– Simulink Control Design
slcontrol\slctrlguis	– (No table of contents file)
slcontrol\slctrlutil	– (No table of contents file)
slcontrol\slctrldemos	– (No table of contents file)
slestim\slestdemos	– Simulink Parameter Estimation Demos
slestim\slestguis	– (No table of contents file)
slestim\slestim	– Simulink Parameter Estimation

slestim\slestmex	– Simulink Parameter Estimation S-Function MEX-files.
slestim\slestutil	– (No table of contents file)
sloptim\sloptim	– Simulink Response Optimization
sloptim\sloptguis	– (No table of contents file)
sloptim\sloptdemos	– Simulink Response Optimization Demos.
sloptim\sloptobsolete	– (No table of contents file)
simulink\slvnv	– Simulink Verification and Validation
simulink\simcoverage	– (No table of contents file)
simcoverage\simcovdemos	– (No table of contents file)
toolbox\splines	– Spline Toolbox
toolbox\stats	– Statistics Toolbox
toolbox\symbolic	– Symbolic Math Toolbox
tic2000\tic2000	– Embedded Target for TI C2000 DSP(tm)
tic2000\tic2000blks	– (No table of contents file)
tic2000\tic2000demos	– (No table of contents file)
etargets\etargets	– (No table of contents file)
tic6000\tic6000	– Embedded Target for TI C6000 DSP(tm)
tic6000\tic6000blks	– TI C6000 (tm) Blocks
tic6000\tic6000demos	– Embedded Target for TI C6000 DSP(tm) Demos
etargets\rtdxblks	– RTDX (tm) Blocks
vipblks\vipblks	– Video and Image Processing Blockset
vipblks\vipmasks	– (No table of contents file)
vipblks\vipmex	– (No table of contents file)
vipblks\vipdemos	– Video and Image Processing Blockset demonstrations and examples.
vr\vr	– Virtual Reality Toolbox
vr\vrdemos	– Virtual Reality Toolbox examples.
wavelet\wavelet	– Wavelet Toolbox
wavelet\wavedemo	– Wavelet Toolbox Demonstrations.

For help on your specific MATLAB installation, type

$\gg$ **help matlab/general**

You will be given information on your software version as follows.

General purpose commands.
  MATLAB Version 7.3 (R2006b) 03-Aug-2006

General information.

syntax	– Help on MATLAB command syntax.
demo	– Run demonstrations.
ver	– MATLAB, Simulink and toolbox version information.

version	– MATLAB version information.

Managing the workspace.

who	– List current variables.
whos	– List current variables, long form.
clear	– Clear variables and functions from memory.
pack	– Consolidate workspace memory.
load	– Load workspace variables from disk.
save	– Save workspace variables to disk.
saveas	– Save Figure or model to desired output format.
memory	– Help for memory limitations.
recycle	– Set option to move deleted files to recycle folder.
quit	– Quit MATLAB session.
exit	– Exit from MATLAB.

Managing commands and functions.

what	– List MATLAB-specific files in directory.
type	– List M-file.
open	– Open files by extension.
which	– Locate functions and files.
pcode	– Create pre-parsed pseudo-code file (P-file).
mex	– Compile MEX-function.
inmem	– List functions in memory.
namelengthmax	– Maximum length of MATLAB function or variable name.

Managing the search path.

path	– Get/set search path.
addpath	– Add directory to search path.
rmpath	– Remove directory from search path.
rehash	– Refresh function and file system caches.
import	– Import Java packages into the current scope.
finfo	– Identify file type against standard file handlers on path.
genpath	– Generate recursive toolbox path.
savepath	– Save the current MATLAB path in the pathdef.m file.

Managing the java search path.

javaaddpath	– Add directories to the dynamic java path.
javaclasspath	– Get and set java path.
javarmpath	– Remove directory from dynamic java path.

Controlling the command window.

echo	– Echo commands in M-files.
more	– Control paged output in command window.

diary	– Save text of MATLAB session.
format	– Set output format.
beep	– Produce beep sound.
desktop	– Start and query the MATLAB Desktop.
preferences	– Bring up MATLAB user settable preferences dialog.

Operating system commands.

cd	– Change current working directory.
copyfile	– Copy file or directory.
movefile	– Move file or directory.
delete	– Delete file or graphics object.
pwd	– Show (print) current working directory.
dir	– List directory.
ls	– List directory.
fileattrib	– Set or get attributes of files and directories.
isdir	– True if argument is a directory.
mkdir	– Make new directory.
rmdir	– Remove directory.
getenv	– Get environment variable.
!	– Execute operating system command (see PUNCT).
dos	– Execute DOS command and return result.
unix	– Execute UNIX command and return result.
system	– Execute system command and return result.
perl	– Execute Perl command and return the result.
computer	– Computer type.
isunix	– True for the UNIX version of MATLAB.
ispc	– True for the PC (Windows) version of MATLAB.

Debugging.

| debug | – List debugging commands. |
| mexdebug | – Debug MEX-files. |

Tools to locate dependent functions of an M-file.

| depfun | – Locate dependent functions of an M-file or P-file. |
| depdir | – Locate dependent directories of an M-file or P-file. |

Loading and calling shared libraries.

calllib	– Call a function in an external library.
libpointer	– Creates a pointer object for use with external libraries.
libstruct	– Creates a structure pointer for use with external libraries.
libisloaded	– True if the specified shared library is loaded.
loadlibrary	– Load a shared library into MATLAB.

libfunctions	– Return information on functions in an external library.
libfunctionsview	– View the functions in an external library.
unloadlibrary	– Unload a shared library loaded with LOADLIBRARY.
java	– Using Java from within MATLAB.
usejava	– True if the specified Java feature is supported in MATLAB.

See also lang, datatypes, iofun, graphics, ops, strfun, timefun, matfun, demos, graphics, datafun, uitools, doc, punct, arith.

The toolboxes and other utilities you have available in your MATLAB installation can be accessed using the command

    ≫ **path**

MATLABPATH

```
C:\Program Files\MATLAB\R2006b\toolbox\matlab\general
C:\Program Files\MATLAB\R2006b\toolbox\matlab\ops
C:\Program Files\MATLAB\R2006b\toolbox\matlab\lang
C:\Program Files\MATLAB\R2006b\toolbox\matlab\elmat
C:\Program Files\MATLAB\R2006b\toolbox\matlab\elfun
C:\Program Files\MATLAB\R2006b\toolbox\matlab\specfun
C:\Program Files\MATLAB\R2006b\toolbox\matlab\matfun
C:\Program Files\MATLAB\R2006b\toolbox\matlab\datafun
C:\Program Files\MATLAB\R2006b\toolbox\matlab\polyfun
C:\Program Files\MATLAB\R2006b\toolbox\matlab\funfun
C:\Program Files\MATLAB\R2006b\toolbox\matlab\sparfun
C:\Program Files\MATLAB\R2006b\toolbox\matlab\scribe
C:\Program Files\MATLAB\R2006b\toolbox\matlab\graph2d
C:\Program Files\MATLAB\R2006b\toolbox\matlab\graph3d
C:\Program Files\MATLAB\R2006b\toolbox\matlab\specgraph
C:\Program Files\MATLAB\R2006b\toolbox\matlab\graphics
C:\Program Files\MATLAB\R2006b\toolbox\matlab\uitools
C:\Program Files\MATLAB\R2006b\toolbox\matlab\strfun
C:\Program Files\MATLAB\R2006b\toolbox\matlab\imagesci
C:\Program Files\MATLAB\R2006b\toolbox\matlab\iofun
C:\Program Files\MATLAB\R2006b\toolbox\matlab\audiovideo
C:\Program Files\MATLAB\R2006b\toolbox\matlab\timefun
C:\Program Files\MATLAB\R2006b\toolbox\matlab\datatypes
C:\Program Files\MATLAB\R2006b\toolbox\matlab\verctrl
C:\Program Files\MATLAB\R2006b\toolbox\matlab\codetools
C:\Program Files\MATLAB\R2006b\toolbox\matlab\helptools
C:\Program Files\MATLAB\R2006b\toolbox\matlab\winfun
C:\Program Files\MATLAB\R2006b\toolbox\matlab\demos
C:\Program Files\MATLAB\R2006b\toolbox\matlab\timeseries
C:\Program Files\MATLAB\R2006b\toolbox\matlab\hds
```

C:\Program Files\MATLAB\R2006b\toolbox\local
C:\Program Files\MATLAB\R2006b\toolbox\shared\controllib
C:\Program Files\MATLAB\R2006b\toolbox\simulink\simulink
C:\Program Files\MATLAB\R2006b\toolbox\simulink\blocks
C:\Program Files\MATLAB\R2006b\toolbox\simulink\components
C:\Program Files\MATLAB\R2006b\toolbox\simulink\fixedandfloat
C:\Program Files\MATLAB\R2006b\toolbox\simulink\fixedandfloat\
fxpdemos
C:\Program Files\MATLAB\R2006b\toolbox\simulink\fixedandfloat\
obsolete
C:\Program Files\MATLAB\R2006b\toolbox\simulink\simdemos
C:\Program Files\MATLAB\R2006b\toolbox\simulink\simdemos\
aerospace
C:\Program Files\MATLAB\R2006b\toolbox\simulink\simdemos\
automotive
C:\Program Files\MATLAB\R2006b\toolbox\simulink\simdemos\
simfeatures
C:\Program Files\MATLAB\R2006b\toolbox\simulink\simdemos\
simgeneral
C:\Program Files\MATLAB\R2006b\toolbox\simulink\dee
C:\Program Files\MATLAB\R2006b\toolbox\shared\dastudio
C:\Program Files\MATLAB\R2006b\toolbox\stateflow\stateflow
C:\Program Files\MATLAB\R2006b\toolbox\rtw\rtw
C:\Program Files\MATLAB\R2006b\toolbox\simulink\simulink\
modeladvisor
C:\Program Files\MATLAB\R2006b\toolbox\simulink\simulink\
modeladvisor\fixpt
C:\Program Files\MATLAB\R2006b\toolbox\simulink\simulink\MPlayIO
C:\Program Files\MATLAB\R2006b\toolbox\simulink\simulink\
dataobjectwizard
C:\Program Files\MATLAB\R2006b\toolbox\shared\fixedpointlib
C:\Program Files\MATLAB\R2006b\toolbox\stateflow\sfdemos
C:\Program Files\MATLAB\R2006b\toolbox\stateflow\coder
C:\Program Files\MATLAB\R2006b\toolbox\rtw\rtwdemos
C:\Program Files\MATLAB\R2006b\toolbox\rtw\rtwdemos\rsimdemos
C:\Program Files\MATLAB\R2006b\toolbox\rtw\targets\asap2\asap2
C:\Program Files\MATLAB\R2006b\toolbox\rtw\targets\asap2\asap2\user
C:\Program Files\MATLAB\R2006b\toolbox\rtw\targets\common\
can\blocks
C:\Program Files\MATLAB\R2006b\toolbox\rtw\targets\common\
configuration\resource
C:\Program Files\MATLAB\R2006b\toolbox\rtw\targets\common\
tgtcommon
C:\Program Files\MATLAB\R2006b\toolbox\rtw\targets\rtwin\rtwin
C:\Program Files\MATLAB\R2006b\toolbox\simulink\accelerator
C:\Program Files\MATLAB\R2006b\toolbox\simulink\accelerator\
acceldemos
C:\Program Files\MATLAB\R2006b\toolbox\rtw\accel

C:\Program Files\MATLAB\R2006b\toolbox\aeroblks\aeroblks
C:\Program Files\MATLAB\R2006b\toolbox\aeroblks\aerodemos
C:\Program Files\MATLAB\R2006b\toolbox\aeroblks\aerodemos\texture
C:\Program Files\MATLAB\R2006b\toolbox\bioinfo\bioinfo
C:\Program Files\MATLAB\R2006b\toolbox\bioinfo\biolearning
C:\Program Files\MATLAB\R2006b\toolbox\bioinfo\microarray
C:\Program Files\MATLAB\R2006b\toolbox\bioinfo\mass_spec
C:\Program Files\MATLAB\R2006b\toolbox\bioinfo\proteins
C:\Program Files\MATLAB\R2006b\toolbox\bioinfo\biomatrices
C:\Program Files\MATLAB\R2006b\toolbox\bioinfo\biodemos
C:\Program Files\MATLAB\R2006b\toolbox\rtw\targets\c166\c166
C:\Program Files\MATLAB\R2006b\toolbox\rtw\targets\cl66\blocks
C:\Program Files\MATLAB\R2006b\toolbox\rtw\targets\cl66\cl66demos
C:\Program Files\MATLAB\R2006b\toolbox\ccslink\ccslink
C:\Program Files\MATLAB\R2006b\toolbox\ccslink\ccslink_outproc
C:\Program Files\MATLAB\R2006b\toolbox\ccslink\ccsblks
C:\Program Files\MATLAB\R2006b\toolbox\ccslink\ccsdemos
C:\Program Files\MATLAB\R2006b\toolbox\comm\comm
C:\Program Files\MATLAB\R2006b\toolbox\comm\commdemos
C:\Program Files\MATLAB\R2006b\toolbox\comm\commdemos\
commdocdemos
C:\Program Files\MATLAB\R2006b\toolbox\comm\commobsolete
C:\Program Files\N4ATLAB\R2006b\toolbox\commblks\commblks
C:\Program Files\MATLAB\R2006b\toolbox\commblks\commmasks
C:\Program Files\MATLAB\R2006b\toolbox\commblks\commmex
C:\Program Files\MATLAB\R2006b\toolbox\commblks\commblksdemos
C:\Program Files\MATLAB\R2006b\toolbox\commblks\commblks
obsolete\v3
C:\Program Files\MATLAB\R2006b\toolbox\commblks\commblks
obsolete\v2p5
C:\Program Files\MATLAB\R2006b\toolbox\commblks\commblks
obsolete\v2
C:\Program Files\MATLAB\R2006b\toolbox\commblks\commblks
obsolete\v1p5
C:\Program Files\MATLAB\R2006b\toolbox\control\control
C:\Program Files\MATLAB\R2006b\toolbox\control\ctrlguis
C:\Program Files\MATLAB\R2006b\toolbox\control\ctrlobsolete
C:\Program Files\MATLAB\R2006b\toolbox\control\ctrlutil
C:\Program Files\MATLAB\R2006b\toolbox\control\ctrldemos
C:\Program Files\MATLAB\R2006b\toolbox\shared\slcontrollib
C:\Program Files\MATLAB\R2006b\toolbox\curvefit\curvefit
C:\Program Files\MATLAB\R2006b\toolbox\curvefit\cftoolgui
C:\Program Files\MATLAB\R2006b\toolbox\shared\optimlib
C:\Program Files\MATLAB\R2006b\toolbox\daq\daq
C:\Program Files\MATLAB\R2006b\toolbox\daq\daqguis
C:\Program Files\MATLAB\R2006b\toolbox\daq\daqdemos
C:\Program Files\MATLAB\R2006b\toolbox\database\database
C:\Program Files\MATLAB\R2006b\toolbox\database\dbdemos

C:\Program Files\MATLAB\R2006b\toolbox\database\vqb
C:\Program Files\MATLAB\R2006b\toolbox\datafeed\datafeed
C:\Program Files\MATLAB\R2006b\toolbox\datafeed\dfgui
C:\Program Files\MATLAB\R2006b\toolbox\des\desblks
C:\Program Files\MATLAB\R2006b\toolbox\des\desmasks
C:\Program Files\MATLAB\R2006b\toolbox\des\desmex
C:\Program Files\MATLAB\R2006b\toolbox\des\desdemos
C:\Program Files\MATLAB\R2006b\toolbox\physmod\drive\drive
C:\Program Files\MATLAB\R2006b\toolbox\physmod\drive\drivedemos
C:\Program Files\MATLAB\R2006b\toolbox\dspblks\dspblks
C:\Program Files\MATLAB\R2006b\toolbox\dspblks\dspmasks
C:\Program Files\MATLAB\R2006b\toolbox\dspblks\dspmex
C:\Program Files\MATLAB\R2006b\toolbox\dspblks\dspdemos
C:\Program Files\MATLAB\R2006b\toolbox\rtw\targets\ecoder
C:\Program Files\MATLAB\R2006b\toolbox\rtw\targets\ecoder\
   ecoderdemos
C:\Program Files\MATLAB\R2006b\toolbox\rtw\targets\mpt
C:\Program Files\MATLAB\R2006b\toolbox\rtw\targets\mpt\mpt
C:\Program Files\MATLAB\R2006b\toolbox\rtw\targets\mpt\
   user_specific
C:\Program Files\MATLAB\R2006b\toolbox\exlink
C:\Program Files\MATLAB\R2006b\toolbox\symbolic\extended
C:\Program Files\MATLAB\R2006b\toolbox\filterdesign\filterdesign
C:\Program Files\MATLAB\R2006b\toolbox\filterdesign\quantization
C:\Program Files\MATLAB\R2006b\toolbox\filterdesign\filtdesdemos
C:\Program Files\MATLAB\R2006b\toolbox\finance\finance
C:\Program Files\MATLAB\R2006b\toolbox\finance\calendar
C:\Program Files\MATLAB\R2006b\toolbox\finance\findemos
C:\Program Files\MATLAB\R2006b\toolbox\finance\finsupport
C:\Program Files\MATLAB\R2006b\toolbox\finance\ftseries
C:\Program Files\MATLAB\R2006b\toolbox\finance\ftsdemos
C:\Program Files\MATLAB\R2006b\toolbox\finance\ftsdata
C:\Program Files\MATLAB\R2006b\toolbox\finance\ftstutorials
C:\Program Files\MATLAB\R2006b\toolbox\finderiv\finderiv
C:\Program Files\MATLAB\R2006b\toolbox\finfixed\finfixed
C:\Program Files\MATLAB\R2006b\toolbox\fixedpoint\fixedpoint
C:\Program Files\MATLAB\R2006b\toolbox\fixedpoint\fidemos
C:\Program Files\MATLAB\R2006b\toolbox\fixedpoint\fimex
C:\Program Files\MATLAB\R2006b\toolbox\fixpoint
C:\Program Files\MATLAB\R2006b\toolbox\fuzzy\fuzzy
C:\Program Files\MATLAB\R2006b\toolbox\fuzzy\fuzdemos
C:\Program Files\MATLAB\R2006b\toolbox\gads
C:\Program Files\MATLAB\R2006b\toolbox\gads\gads
C:\Program Files\MATLAB\R2006b\toolbox\gads\gadsdemos
C:\Program Files\MATLAB\R2006b\toolbox\garch\garch
C:\Program Files\MATLAB\R2006b\toolbox\garch\garchdemos
C:\Program Files\MATLAB\R2006b\toolbox\gauges
C:\Program Files\MATLAB\R2006b\toolbox\rtw\targets\hc12\hc12

```
C:\Program Files\MATLAB\R2006b\toolbox\rtw\targets\hc12\blocks
C:\Program Files\MATLAB\R2006b\toolbox\rtw\targets\hc12\codewarrior
C:\Program Files\MATLAB\R2006b\toolbox\rtw\targets\hc12\hc12demos
C:\Program Files\MATLAB\R2006b\toolbox\hdlfilter\hdlfilter
C:\Program Files\MATLAB\R2006b\toolbox\hdlfilter\hdlfiltdemos
C:\Program Files\MATLAB\R2006b\toolbox\shared\hdlshared
C:\Program Files\MATLAB\R2006b\toolbox\ident\ident
C:\Program Files\MATLAB\R2006b\toolbox\ident\idobsolete
C:\Program Files\MATLAB\R2006b\toolbox\ident\idguis
C:\Program Files\MATLAB\R2006b\toolbox\ident\idutils
C:\Program Files\MATLAB\R2006b\toolbox\ident\iddemos
C:\Program Files\MATLAB\R2006b\toolbox\ident\idhelp
C:\Program Files\MATLAB\R2006b\toolbox\images\images
C:\Program Files\MATLAB\R2006b\toolbox\images\imuitools
C:\Program Files\MATLAB\R2006b\toolbox\images\imdemos
C:\Program Files\MATLAB\R2006b\toolbox\images\iptutils
C:\Program Files\MATLAB\R2006b\toolbox\shared\imageslib
C:\Program Files\MATLAB\R2006b\toolbox\images\medformats
C:\Program Files\MATLAB\R2006b\toolbox\imaq\imaq
C:\Program Files\MATLAB\R2006b\toolbox\shared\imaqlib
C:\Program Files\MATLAB\R2006b\toolbox\imaq\imaqdemos
C:\Program Files\MATLAB\R2006b\toolbox\imaq\imaqblks\imaqblks
C:\Program Files\MATLAB\R2006b\toolbox\imaq\imaqblks\imaqmasks
C:\Program Files\MATLAB\R2006b\toolbox\imaq\imaqblks\imaqmex
C:\Program Files\MATLAB\R2006b\toolbox\instrument\instrument
C:\Program Files\MATLAB\R2006b\toolbox\instrument\instrumentdemos
C:\Program Files\MATLAB\R2006b\toolbox\instrument\instrumentblks\
   instrumentblks
C:\Program Files\MATLAB\R2006b\toolbox\instrument\instrumentblks\
   instrumentmex
C:\Program Files\MATLAB\R2006b\toolbox\map\map
C:\Program Files\MATLAB\R2006b\toolbox\map\mapdemos
C:\Program Files\MATLAB\R2006b\toolbox\map\mapdisp
C:\Program Files\MATLAB\R2006b\toolbox\map\mapformats
C:\Program Files\MATLAB\R2006b\toolbox\map\mapproj
C:\Program Files\MATLAB\R2006b\toolbox\shared\mapgeodesy
C:\Program Files\MATLAB\R2006b\toolbox\mbc\mbc
C:\Program Files\MATLAB\R2006b\toolbox\mbc\mbcdata
C:\Program Files\MATLAB\R2006b\toolbox\mbc\mbcdesign
C:\Program Files\MATLAB\R2006b\toolbox\mbc\mbcexpr
C:\Program Files\MATLAB\R2006b\toolbox\mbc\mbcguitools
C:\Program Files\MATLAB\R2006b\toolbox\mbc\mbclayouts
C:\Program Files\MATLAB\R2006b\toolbox\mbc\mbcmodels
C:\Program Files\MATLAB\R2006b\toolbox\mbc\mbcsimulink
C:\Program Files\MATLAB\R2006b\toolbox\mbc\mbctools
C:\Program Files\MATLAB\R2006b\toolbox\mbc\mbcview
C:\Program Files\MATLAB\R2006b\toolbox\physmod\mech\mech
C:\Program Files\MATLAB\R2006b\toolbox\physmod\mech\mechdemos
```

C:\Program Files\MATLAB\R2006b\toolbox\physmod\pmimport\
pmimport
C:\Program Files\MATLAB\R2006b\toolbox\slvnv\simcoverage
C:\Program Files\MATLAB\R2006b\toolbox\modelsim\modelsim
C:\Program Files\MATLAB\R2006b\toolbox\modelsim\modelsimdemos
C:\Program Files\MATLAB\R2006b\toolbox\mpc\mpc
C:\Program Files\MATLAB\R2006b\toolbox\mpc\mpcdemos
C:\Program Files\MATLAB\R2006b\toolbox\mpc\mpcguis
C:\Program Files\MATLAB\R2006b\toolbox\mpc\mpcobsolete
C:\Program Files\MATLAB\R2006b\toolbox\mpc\mpcutils
C:\Program Files\MATLAB\R2006b\toolbox\rtw\targets\mpc555dk
C:\Program Files\MATLAB\R2006b\toolbox\rtw\targets\mpc555dk\
common\configuration
C:\Program Files\MATLAB\R2006b\toolbox\rtw\targets\mpc555dk\
mpc555demos
C:\Program Files\MATLAB\R2006b\toolbox\rtw\targets\mpc555dk\
mpc555dk
C:\Program Files\MATLAB\R2006b\toolbox\rtw\targets\mpc555dk\pil
C:\Program Files\MATLAB\R2006b\toolbox\rtw\targets\mpc555dk\rt\
blockset\mfiles
C:\Program Files\MATLAB\R2006b\toolbox\rtw\targets\mpc555dk\rt\
blockset
C:\Program Files\MATLAB\R2006b\toolbox\nnet
C:\Program Files\MATLAB\R2006b\toolbox\nnet\nncontrol
C:\Program Files\MATLAB\R2006b\toolbox\nnet\nndemos
C:\Program Files\MATLAB\R2006b\toolbox\nnet\nnet
C:\Program Files\MATLAB\R2006b\toolbox\nnet\nnet\nnanalyze
C:\Program Files\MATLAB\R2006b\toolbox\nnet\nnet\nncustom
C:\Program Files\MATLAB\R2006b\toolbox\nnet\nnet\nndistance
C:\Program Files\MATLAB\R2006b\toolbox\nnet\nnet\nnformat
C:\Program Files\MATLAB\R2006b\toolbox\nnet\nnet\nninit
C:\Program Files\MATLAB\R2006b\toolbox\nnet\nnet\nnlearn
C:\Program Files\MATLAB\R2006b\toolbox\nnet\nnet\nnnetinput
C:\Program Files\MATLAB\R2006b\toolbox\nnet\nnet\nnnetwork
C:\Program Files\MATLAB\R2006b\toolbox\nnet\nnet\nnperformance
C:\Program Files\MATLAB\R2006b\toolbox\nnet\nnet\nnplot
C:\Program Files\MATLAB\R2006b\toolbox\nnet\nnet\nnprocess
C:\Program Files\MATLAB\R2006b\toolbox\nnet\nnet\nnsearch
C:\Program Files\MATLAB\R2006b\toolbox\nnet\nnet\nntopology
C:\Program Files\MATLAB\R2006b\toolbox\nnet\nnet\nntrain
C:\Program Files\MATLAB\R2006b\toolbox\nnet\nnet\nntransfer
C:\Program Files\MATLAB\R2006b\toolbox\nnet\nnet\nnweight
C:\Program Files\MATLAB\R2006b\toolbox\nnet\nnguis
C:\Program Files\MATLAB\R2006b\toolbox\nnet\nnguis\nftool
C:\Program Files\MATLAB\R2006b\toolbox\nnet\nnguis\nntool
C:\Program Files\MATLAB\R2006b\toolbox\nnet\nnobsolete
C:\Program Files\MATLAB\R2006b\toolbox\nnet\nnresource
C:\Program Files\MATLAB\R2006b\toolbox\nnet\nnutils

C:\Program Files\MATLAB\R2006b\toolbox\opc\opc
C:\Program Files\MATLAB\R2006b\toolbox\opc\opcgui
C:\Program Files\MATLAB\R2006b\toolbox\opc\opcdemos
C:\Program Files\MATLAB\R2006b\toolbox\opc\opcdemos\opcblksdemos
C:\Program Files\MATLAB\R2006b\toolbox\opc\opcblks\opcblks
C:\Program Files\MATLAB\R2006b\toolbox\opc\opcblks\opcmasks
C:\Program Files\MATLAB\R2006b\toolbox\optim
C:\Program Files\MATLAB\R2006b\toolbox\rtw\targets\osek\osek
C:\Program Files\MATLAB\R2006b\toolbox\rtw\targets\osek\osekdemos
C:\Program Files\MATLAB\R2006b\toolbox\rtw\targets\osek\blocks
C:\Program Files\MATLAB\R2006b\toolbox\rtw\targets\osek\osekworks
C:\Program Files\MATLAB\R2006b\toolbox\rtw\targets\osek\proosek
C:\Program Files\MATLAB\R2006b\toolbox\pde
C:\Program Files\MATLAB\R2006b\toolbox\physmod\pm_util\pm_util
C:\Program Files\MATLAB\R2006b\toolbox\physmod\powersys\powersys
C:\Program Files\MATLAB\R2006b\toolbox\physmod\powersys\
powerdemo
C:\Program Files\MATLAB\R2006b\toolbox\physmod\powersys\drives\
drives
C:\Program Files\MATLAB\R2006b\toolbox\physmod\powersys\drives
\drivesdemo
C:\Program Files\MATLAB\R2006b\toolbox\physmod\powersys\facts\facts
C:\Program Files\MATLAB\R2006b\toolbox\physmod\powersys\facts
\factsdemo
C:\Program Files\MATLAB\R2006b\toolbox\physmod\powersys\DR\DR
C:\Program Files\MATLAB\R2006b\toolbox\physmod\powersys\DR\
DRdemo
C:\Program Files\MATLAB\R2006b\toolbox\slvnv\reqmgt
C:\Program Files\MATLAB\R2006b\toolbox\slvnv\rmidemos
C:\Program Files\MATLAB\R2006b\toolbox\rf\rf
C:\Program Files\MATLAB\R2006b\toolbox\rf\rfdemos
C:\Program Files\MATLAB\R2006b\toolbox\rf\rftool
C:\Program Files\MATLAB\R2006b\toolbox\rfblks\rfblks
C:\Program Files\MATLAB\R2006b\toolbox\rfblks\rfblksmasks
C:\Program Files\MATLAB\R2006b\toolbox\rfblks\rfblksmex
C:\Program Files\MATLAB\R2006b\toolbox\rfblks\rfblksdemos
C:\Program Files\MATLAB\R2006b\toolbox\robust\robust
C:\Program Files\MATLAB\R2006b\toolbox\robust\rctlmi
C:\Program Fies\MATLAB\R2006b\toolbox\robust\rctutil
C:\Program Files\MATLAB\R2006b\toolbox\robust\rctdemos
C:\Program Files\MATLAB\R2006b\toolbox\robust\rctobsolete\robust
C:\Program Files\MATLAB\R2006b\toolbox\robust\rctobsolete\lmi
C:\Program Files\MATLAB\R2006b\toolbox\robust\rctobsolete\mutools\
commands
C:\Program Files\MATLAB\R2006b\toolbox\robust\rctobsolete\
mutools\subs
C:\Program Files\MATLAB\R2006b\toolbox\rptgen\rptgen
C:\Program Files\MATLAB\R2006b\toolbox\rptgen\rptgendemos

C:\Program Files\MATLAB\R2006b\toolbox\rptgen\rptgenv1
C:\Program Files\MATLAB\R2006b\toolbox\rptgenext\rptgenext
C:\Program Files\MATLAB\R2006b\toolbox\xptgenext\rptgenextdemos
C:\Program Files\MATLAB\R2006b\toolbox\rptgenext\rptgenextv1
C:\Program Files\MATLAB\R2006b\toolbox\signal\signal
C:\Program Files\MATLAB\R2006b\toolbox\signal\sigtools
C:\Program Files\MATLAB\R2006b\toolbox\signal\sptoolgui
C:\Program Files\MATLAB\R2006b\toolbox\signal\sigdemos
C:\Program Files\MATLAB\R2006b\toolbox\simbio\simbio
C:\Program Files\MATLAB\R2006b\toolbox\simbio\simbiodemos
C:\Program Files\MATLAB\R2006b\toolbox\slcontrol\s1control
C:\Program Files\MATLAB\R2006b\toolbox\slcontrol\slctrlguis
C:\Program Files\MATLAB\R2006b\toolbox\slcontrol\slctrlutil
C:\Program Files\MATLAB\R2006b\toolbox\slcontrol\slctrldemos
C:\Program Files\MATLAB\R2006b\toolbox\slestim\slestdemos
C:\Program Files\MATLAB\R2006b\toolbox\slestim\slestguis
C:\Program Files\MATLAB\R2006b\toolbox\slestim\slestim
C:\Program Files\MATLAB\R2006b\toolbox\slestim\slestmex
C:\Program Files\MATLAB\R2006b\toolbox\slestim\slestutil
C:\Program Files\MATLAB\R2006b\toolbox\sloptim\sloptim
C:\Program Files\MATLAB\R2006b\toolbox\sloptim\sloptguis
C:\Program Files\MATLAB\R2006b\toolbox\sloptim\sloptdemos
C:\Program Files\MATLAB\R2006b\toolbox\sloptim\sloptobsolete
C:\Program Files\MATLAB\R2006b\toolbox\slvnv\slvnv
C:\Program Files\MATLAB\R2006b\toolbox\slvnv\simcovdemos
C:\Program Files\MATLAB\R2006b\toolbox\splines
C:\Program Files\MATLAB\R2006b\toolbox\stats
C:\Program Files\MATLAB\R2006b\toolbox\symbolic
C:\Program Files\MATLAB\R2006b\toolbox\rtw\targets\tic2000\
tic2000
C:\Program Files\MATLAB\R2006b\toolbox\rtw\targets\tic2000\
tic2000blks
C:\Program Files\MATLAB\R2006b\toolbox\rtw\targets\tic2000\
tic2000demos
C:\Program Files\MATLAB\R2006b\toolbox\shared\etargets\etargets
C:\Program Files\MATLAB\R2006b\toolbox\shared\etargets\rtdxblks
C:\Program Files\MATLAB\R2006b\toolbox\rtw\targets\tic6000\tic6000
C:\Program Files\MATLAB\R2006b\toolbox\rtw\targets\tic6000\
tic6000blks
C:\Program Files\MATLAB\R2006b\toolbox\rtw\targets\tic6000\
tic6000demos
C:\Program Files\MATLAB\R2006b\toolbox\vipblks\vipblks
C:\Program Files\MATLAB\R2006b\toolbox\vipblks\vipmasks
C:\Program Files\MATLAB\R2006b\toolbox\vipblks\vipmex
C:\Program Files\MATLAB\R2006b\toolbox\vipblks\vipdemos
C:\Program Files\MATLAB\R2006b\toolbox\vr\vr
C:\Program Files\MATLAB\R2006b\toolbox\vr\vrdemos
C:\Program Files\MATLAB\R2006b\toolbox\wavelet\wavelet

C:\Program Files\MATLAB\R2006b\toolbox\wavelet\wavedemo
C:\Program Files\MATLAB\R2006b\toolbox\rtw\targets\xpc\xpc
C:\Program
   Files\MATLAB\R2006b\toolbox\rtw\targets\xpc\target\build\xpcblocks
C:\Program Files\MATLAB\R2006b\toolbox\rtw\targets\xpc\xpcdemos
C:\Program Files\MATLAB\R2006b\toolbox\rtw\targets\xpc\xpc\
   xpcmngr
C:\Program Files\MATLAB\R2006b\work
C:\Program Files\MATLAB\R2006b\toolbox\physmod\network_engine\
   network_engine
C:\Program
   Files\MATLAB\R2006b\toolbox\physmod\network_engine\ne_sli
C:\Program
   Files\MATLAB\R2006b\toolbox\physmod\network_engine\library
C:\Program Files\MATLAB\R2006b\toolbox\physmod\sh\sh
C:\Program Files\MATLAB\R2006b\toolbox\physmod\sh\shdemos
C:\Program Files\MATLAB\R2006b\toolbox\physmod\sh\library

This is often a more representative and detailed path listing than is given in your **Launch Pad** window.

To find out what version of MATLAB and its toolboxes you are using, type the command

≫ **ver**

It will provide you with version information as follows.

MATLAB Version 7.3.0.267 (R2006b)
MATLAB License Number: DEMO
Operating System: Microsoft Windows XP Version 5.1 (Build 2600: Service Pack 2)
Java VM Version: Java 1.5.0 with Sun Microsystems Inc. Java HotSpot(TM) Client VM
mixed mode

MATLAB	Version 7.3	(R2006b)
Simulink	Version 6.5	(R2006b)
Aerospace Blockset	Version 2.2	(R2006b)
Aerospace Toolbox	Version 1.0	(R2006b)
Bioinformatics Toolbox	Version 2.4	(R2006b)
Communications Blockset	Version 3.4	(R2006b)
Communications Toolbox	Version 3.4	(R2006b)
Control System Toolbox	Version 7.1	(R2006b)
Curve Fitting Toolbox	Version 1.1.6	(R2006b)
Data Acquisition Toolbox	Version 2.9	(R2006b)
Database Toolbox	Version 3.2	(R2006b)
Datafeed Toolbox	Version 1.9	(R2006b)
Embedded Target for Infineon C166 Microcontrollers	Version 1.3	(R2006b)
Embedded Target for Motorola MPC555	Version 2.0.5	(R2006b)

Embedded Target for TI C2000 DSP(tm)	Version 2.1	(R2006b)
Embedded Target for TI C6000 DSP(tm)	Version 3.1	(R2006b)
Excel Link	Version 2.4	(R2006b)
Extended Symbolic Math Toolbox	Version 3.1.5	(R2006b)
Filter Design HDL Coder	Version 1.5	(R2006b)
Filter Design Toolbox	Version 4.0	(R2006b)
Financial Derivatives Toolbox	Version 4.1	(R2006b)
Financial Toolbox	Version 3.1	(R2006b)
Fixed-Income Toolbox	Version 1.2	(R2006b)
Fixed-Point Toolbox	Version 1.5	(R2006b)
Fuzzy Logic Toolbox	Version 2.2.4	(R2006b)
GARCH Toolbox	Version 2.3	(R2006b)
Gauges Blockset	Version 2.0.4	(R2006b)
Genetic Algorithm and Direct Search Toolbox	Version 2.0.2	(R2006b)
Image Acquisition Toolbox	Version 2.0	(R2006b)
Image Processing Toolbox	Version 5.3	(R2006b)
Instrument Control Toolbox	Version 2.4.1	(R2006b)
Link for Code Composer Studio	Version 2.1	(R2006b)
Link for ModelSim	Version 2.1	(R2006b)
Link for TASKING	Version 1.0.1	(R2006b)
MATLAB Report Generator	Version 3.1	(R2006b)
Mapping Toolbox	Version 2.4	(R2006b)
Model Predictive Control Toolbox	Version 2.2.3	(R2006b)
Model-Based Calibration Toolbox	Version 3.1	(R2006b)
Neural Network Toolbox	Version 5.0.1	(R2006b)
OPC Toolbox	Version 2.0.3	(R2006b)
Optimization Toolbox	Version 3.1	(R2006b)
Partial Differential Equation Toolbox	Version 1.0.9	(R2006b)
RF Blockset	Version 1.3.1	(R2006b)
RF Toolbox	Version 2.0	(R2006b)
Real-Time Windows Target	Version 2.6.2	(R2006b)
Real-Time Workshop	Version 6.5	(R2006b)
Real-Time Workshop Embedded Coder	Version 4.5	(R2006b)
Robust Control Toolbox	Version 3.1.1	(R2006b)
Signal Processing Blockset	Version 6.4	(R2006b)
Signal Processing Toolbox	Version 6.6	(R2006b)
SimBiology	Version 2.0.1	(R2006b)
SimDriveline	Version 1.2.1	(R2006b)
SimEvents	Version 1.2	(R2006b)
SimHydraulics	Version 1.1	(R2006b)
SimMechanics	Version 2.5	(R2006b)
SimPowerSystems	Version 4.3	(R2006b)
Simulink Accelerator	Version 6.5	(R2006b)
Simulink Control Design	Version 2.0.1	(R2006b)
Simulink Fixed Point	Version 5.3	(R2006b)
Simulink HDL Coder	Version 1.0	(R2006b)

Simulink Parameter Estimation	Version 1.1.4	(R2006b)
Simulink Report Generator	Version 3.1	(R2006b)
Simulink Response Optimization	Version 3.1	(R2006b)
Simulink Verification and Validation	Version 2.0	(R2006b)
Spline Toolbox	Version 3.3.1	(R2006b)
Stateflow	Version 6.5	(R2006b)
Stateflow Coder	Version 6.5	(R2006b)
Statistics Toolbox	Version 5.3	(R2006b)
Symbolic Math Toolbox	Version 3.1.5	(R2006b)
System Identification Toolbox	Version 6.2	(R2006b)
SystemTest	Version 1.0.1	(R2006b)
Video and Image Processing Blockset	Version 2.2	(R2006b)
Virtual Reality Toolbox	Version 4.4	(R2006b)
Wavelet Toolbox	Version 3.1	(R2006b)
xPC Target	Version 3.1	(R2006b)

If you are interested in all of the M-files and MAT-files you have in your working directory, type the command

≫ **what**

This provides the following information in the MATLAB Command Window:

M-files in the current directory U:\ltimatlb\newltimatlab

ANINTRODUCTION	JETDEMO	POPDEMO
DEMO	ODES	RDEMO
DISKDEMO	PLOTDEMO	WOWPLOT

MAT-files in the current directory U:\ltimatlb\newltimatlab

demo	testing	trial

If you would like contact information for The MathWorks, type the command

≫ **info**

This provides you with the following information:

For information about The MathWorks, go to: http://www.mathworks.com/company/aboutus/contact_us or +508-647-7000.

Other information on MATLAB and The MathWorks is as follows:

MATLAB is available for Windows, Solaris, HP-UX, LINUX, and MacIntosh.

For an up-to-date list of MathWorks Products, visit our Web site at www.mathworks.com.

24 hour access to our Technical Support problem/solution database as well as our FAQ, Technical Notes, and example files is also available at www.mathworks.com.

For MATLAB assistance or information, contact your local representative or:

The MathWorks, Inc.
3 Apple Hill Drive
Natick, MA 01760-2098 USA

Contact Information:
  Phone: +508-647-7000
  Fax: +508-647-7001
  Web: www.mathworks.com
  Newsgroup: comp.soft-sys.matlab
  FTP: ftp.mathworks.com

E-mail:

info@mathworks.com	Sales, pricing, and general information
support@mathworks.com	Technical support for all products
doc@mathworks.com	Documentation error reports
bugs@mathworks.com	Bug reports
service@mathworks.com	Order status, invoice, and license issues
renewals@mathworks.com	Renewal/subscription pricing
pricing@mathworks.com	Product and pricing information
access@mathworks.com	MATLAB Access Program
suggest@mathworks.com	Product enhancement suggestions
news-notes@mathworks.com	MATLAB News & Notes Editor
connections@mathworks.com	MATLAB Connections Program

The MathWorks Web site www.mathworks.com is an excellent resource for MATLAB third-party routines, program demonstrations, and product documentation.

When contacting The MathWorks for license updates and trouble calls, it is very useful to have the Host Identification Number. This is accessed as follows:

>> **hostid**

'999666'

To test the performance of MATLAB on your computer, use the command

>> **bench**

ans =

    0.8493   0.3961   0.3522   0.5232   0.6125   1.8100

This generates a figure and a table comparing your computer with several others as in Figs 1.4 and 1.5.

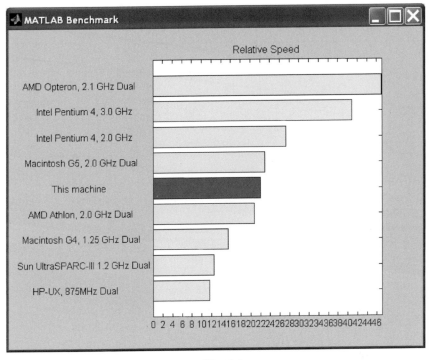

Fig. 1.4

Finally, to see what is new with your version of MATLAB, type the following command:

≫ **whatsnew**

This command brings up the current release notes in the **Help** window. The release notes are also accessible from the **Contents** list on the left-hand

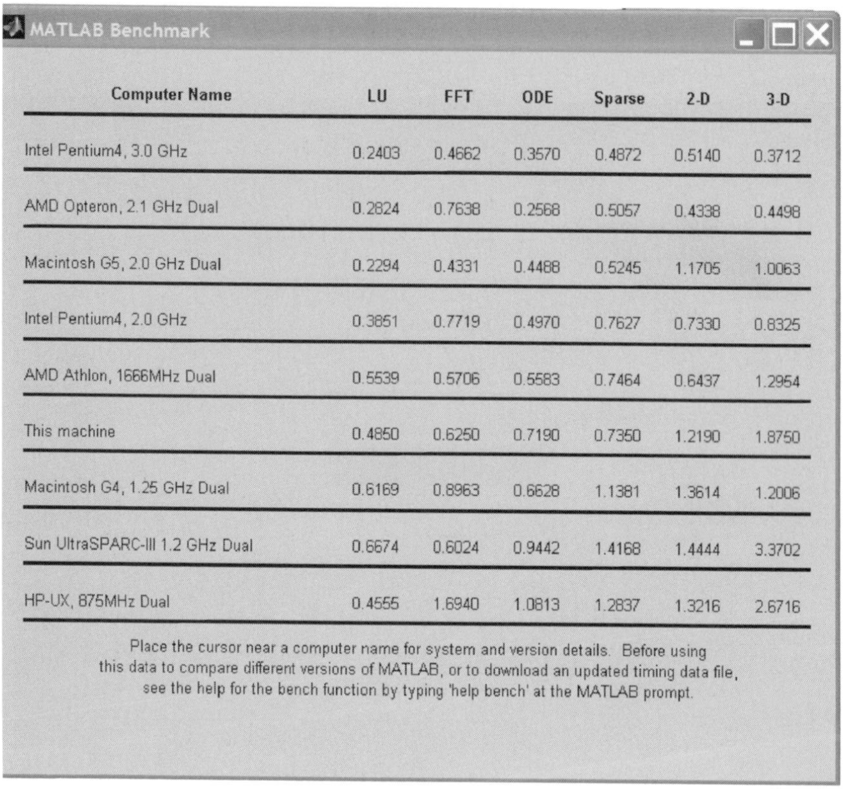

Computer Name	LU	FFT	ODE	Sparse	2-D	3-D
Intel Pentium4, 3.0 GHz	0.2403	0.4662	0.3570	0.4872	0.5140	0.3712
AMD Opteron, 2.1 GHz Dual	0.2824	0.7638	0.2568	0.5057	0.4338	0.4498
Macintosh G5, 2.0 GHz Dual	0.2294	0.4331	0.4488	0.5245	1.1705	1.0063
Intel Pentium4, 2.0 GHz	0.3851	0.7719	0.4970	0.7627	0.7330	0.8325
AMD Athlon, 1666MHz Dual	0.5539	0.5706	0.5583	0.7464	0.6437	1.2954
This machine	0.4850	0.6250	0.7190	0.7350	1.2190	1.8750
Macintosh G4, 1.25 GHz Dual	0.6169	0.8963	0.6628	1.1381	1.3614	1.2006
Sun UltraSPARC-III 1.2 GHz Dual	0.6674	0.6024	0.9442	1.4168	1.4444	3.3702
HP-UX, 875MHz Dual	0.4555	1.6940	1.0813	1.2837	1.3216	2.6716

Place the cursor near a computer name for system and version details. Before using this data to compare different versions of MATLAB, or to download an updated timing data file, see the help for the bench function by typing 'help bench' at the MATLAB prompt.

Fig. 1.5

portion of the **Help** window. The release notes for Release 2006b corresponding to MATLAB version 7.3.0.267 appear in Fig. 1.6.

Try out some of these commands and GUIs. You have just taken the first step in becoming a proficient MATLAB user!

Fig. 1.6

1.11 Conclusion

This chapter introduces the reader to MATLAB and several of the basic matrix computational tools and GUIs that are available in MATLAB through the main **Command Window**.

Practice Exercises

1.1   Enter the following matrix and complete the following computations on this matrix:

a =

0.6000	1.5000	2.3000	−0.5000
8.2000	0.5000	−0.1000	−2.0000
5.7000	8.2000	9.0000	1.5000
0.5000	0.5000	2.4000	0.5000
1.2000	−2.3000	−4.5000	0.5000

a) What is the size of a?
b) Is the matrix square?
c) Which elements of the a matrix are equal to 0.5?
d) Use MATLAB to show the negative matrix elements.
e) b = a(:,2)
f) c = a(4,:)
g) d = [10:15]
h) e = [4:9,1:6]
i) f = [− 5,5]
j) g = [0.0:0.1:1.0]
k) h = a(4:5,1:3)
l) k = a(1:2:5,:)

Notes

2
Plotting and Graphics

2.1 Introduction and Objectives

This chapter covers the plotting and graphics tools that are available in MATLAB®. This includes generating some simple MATLAB animations.

Upon completion of this chapter, the reader will be able to identify and input some basic plotting and graphic tools and commands in MATLAB and generate simple MATLAB movies.

2.2 Plot

The standard linear two-dimensional plot command in MATLAB is **plot(x,y)**, where x and y are vectors. If either **x** or **y** is a matrix, then the vector is plotted versus the rows or columns of the matrix, whichever line up.

Let us first generate a simple sine wave plot. The variable **t** goes from 0 to 10 in increments of 0.3. The variable **y** is the sine of **t**. The semicolon is used to suppress the MATLAB echo.

A title is added to the plot using the command of the same name. A hard copy is generated using the **print** command. Note that the print can also be generated by clicking the printer icon on the plot window, by selecting **Print . . .** from the **File** pull-down menu, or by simultaneously pressing the **Ctrl** and **P** keys with the plot window in the foreground. As with many operations in MATLAB, there are several ways to accomplish the same operation. The best choice depends on the type of information processing being used (i.e., interactive or preprogrammed) and on user preference.

```
≫ t = 0:.3:10;
≫ y = sin(t);
≫ plot(t,y)
≫ title('A simple X-Y plot')
```

43

The resulting plot in the MATLAB plot window appears as shown in Fig. 2.1.

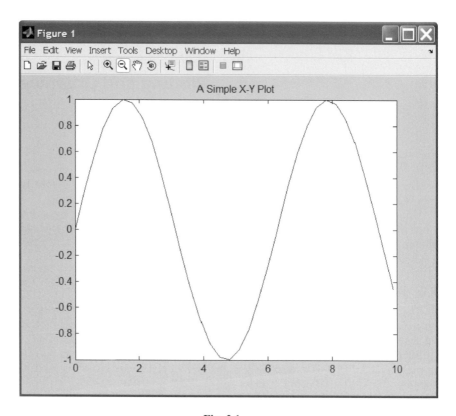

Fig. 2.1

Symbols can be used instead of a solid curve by placing the desired symbol in single quotation marks at the end of the plot command. The **grid** command is used to add grid lines at the major tick marks.

Labels can be placed on the abscissa and the ordinate by using the **xlabel** and the **ylabel** commands. The **gtext('string')** can be used to place a string of text anywhere on the plot. It waits for the mouse button or keyboard key to be pressed while the cursor is within the graphics window. This writes the string onto the graph at the selected location. To see all the possible combinations of colors, markers, and line types, use the command **help plot**.

Let us next modify the previous plot commands to generate a plot with a plus symbol + marking each data point and with grid lines added. Type the following into the MATLAB **Command Window:**

```
>> plot(t,y,' + ')
>> title('Now with a + symbol, and with grid lines')
>> grid
>> xlabel('I do labels too.')
>> ylabel('Hello, World.')
```

The resulting plot from the MATLAB plot window appears as shown in Fig. 2.2.

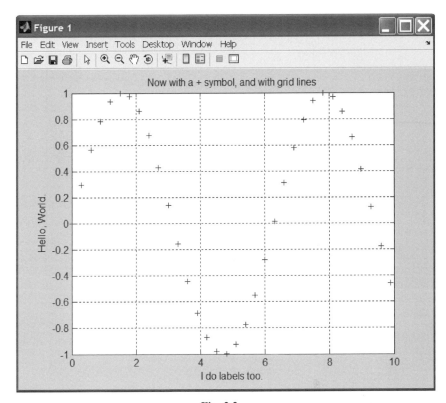

Fig. 2.2

When multiple curves are plotted on the same plot, MATLAB automatically changes the line type to a variety of colors. Older versions of MATLAB use dashed curves (see Fig. 2.3). Multiple curves can be plotted by repeating the x and y vectors within the **plot** command.

```
>> plot(t,y,t,2*y,t,3*y,t,4*y)
>> title('Four different line-types')
```

Marker types can be used to differentiate between various curves. The markers are entered in the plot command after the x-y pair to be plotted using that marker. Note that line types can be combined in the same plot command, as in **plot(x,y,'b-o');**, which plots using blue circle markers (see Fig. 2.4).

```
>> t = 0:.5:10;
>> plot(t,t,'.',t,2*t + 3,' + ',t,3*t + 6,'*',t,4*t + 9,'o',t,5*t + 12,'x')
>> title('Five different marker-types')
```

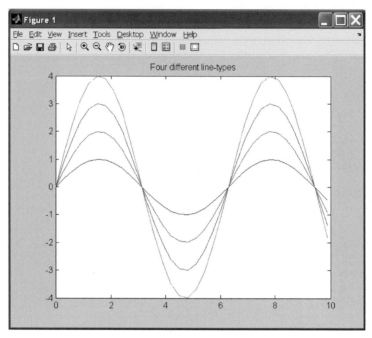

Fig. 2.3

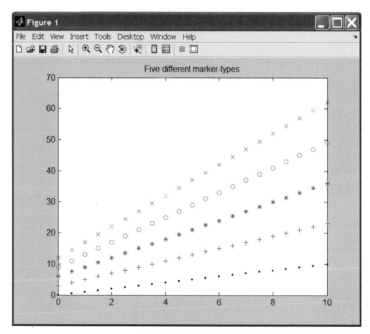

Fig. 2.4

2.3 Log and Semilog Plots

Log scale plots are generated using the **loglog(x,y)** command. The **semilogx(x,y)** command makes a plot using a base 10 logarithmic scale for the x axis and a linear scale for the y axis. The **semilogy(x,y)** command makes a plot using a base 10 logarithmic scale for the y axis and a linear scale for the x axis (see Fig. 2.5). All three operate identically to the plot command. Note that **.*** produces an element-by-element multiplication [t(1)*t(1), t(2)*t(2), . . . , t(n)*t(n)].

> ≫ **t = .1:.1:3;**
> ≫ **loglog(exp(t),exp(t.*t))**
> ≫ **title('I do loglog and semilog plots')**

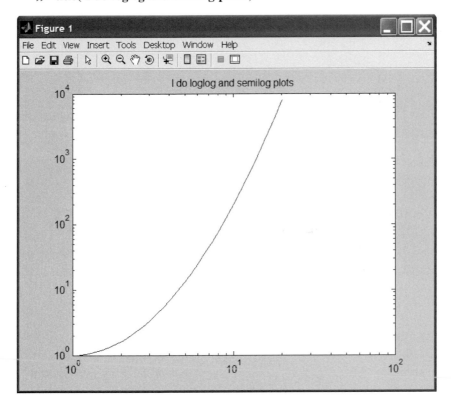

Fig. 2.5

2.4 Polar Plots

The command **polar(theta,rho)** makes a polar coordinate plot of the angle theta (in radians) versus the radius rho (see Fig. 2.6). The line type can be changed by inserting characters within single quotes as in the **plot** command as the third argument. Note that the variable **pi** is predefined by MATLAB to be **3.14159265358979** or **4*atan(1)**.

```
>> t = 0:.05:pi + .1;
>> y = sin(5*t);
>> polar(t,y)
>> title('And polar plots too')
```

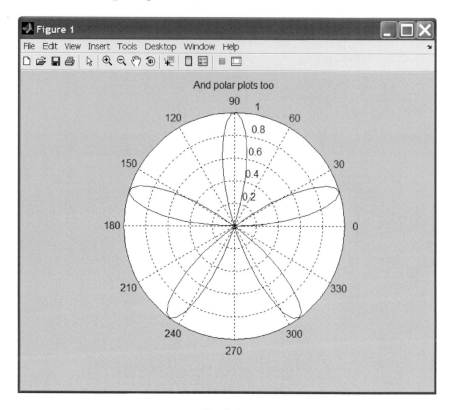

Fig. 2.6

2.5 Subplots

The command **subplot(m,n,p)** breaks the figure window into an m-by-n matrix of small rectangular panes, creates an axis in the pth panel, and makes it current (see Fig. 2.7). The command **subplot(h)** makes the hth axis current. Use **clf** or **clf reset** to return to the default **subplot(1,1,1)** configuration.

Note that the commas can be dropped, modifying the result of the command slightly, and that **subplot = subplot(1,1,1)**.

The command **subplot(111)** without the commas is a special case of **subplot** that does not immediately create an axis. Thus **subplot(111)** is not identical in behavior to **subplot(1,1,1)**.

It sets up the figure so that the next graphics command executes **clf reset** in the figure (deleting all children of the figure) and creates a new axis in the default position. The delayed **clf reset** is accomplished by setting the figure's **NextPlot** to **replace.**

In this example the figure window is divided into a 2-by-2 space. A different plot is placed in each space.

```
≫ t = 0:.3:30;
≫ subplot(221), plot(t,sin(t)),title('Subplots')
≫ subplot(222), plot(t,t.*sin(t))
≫ subplot(223), plot(t,t.*sin(t).^2)
≫ subplot(224), plot(t,t.^2.*sin(t).^2)
≫ subplot
```

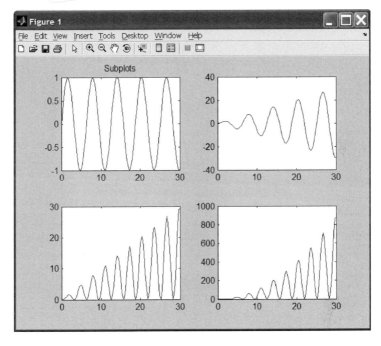

Fig. 2.7

2.6 Axis

Many of the plot attributes can be modified by using the **axis** command. These include using the autoscale mode, entering minimum and maximum axis values, freezing scalings for subsequent plots, turning axis labeling on and off, and setting the axes region to be square (see Fig. 2.8). The command **axis('state')** returns three strings indicating the current settings of the three axis-labeling properties. Note that this can also be accomplished using the plot window's GUI interface.

For example, let us define a vector and then plot a growing cosine wave with the axes squared:

```
≫ t = 0:(.99*pi/2):500;
≫ x = t.*cos(t);
≫ plot(x,t)
≫ axis('square')
```

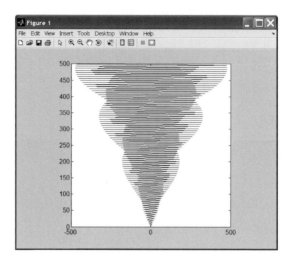

Fig. 2.8

Plotting each point with a period '.' makes the behavior of the cosine wave much clearer.

```
>> plot(x,t,'.')
>> axis('square')
```

The plot resulting from the previous MATLAB commands is given in Fig. 2.9 for comparison purposes.

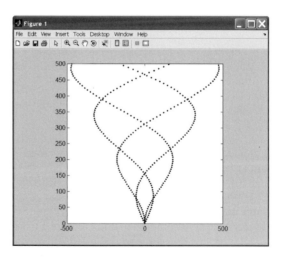

Fig. 2.9

Now let us plot the growing cosine versus a growing sine (see Fig. 2.10):

```
>> y = t.*sin(t);
>> plot(x,y)
>> axis('square')
```

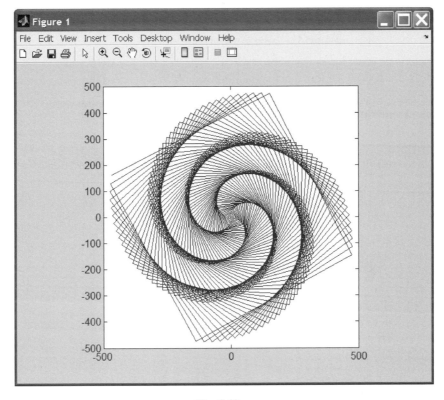

Fig. 2.10

2.7 Mesh

The standard three-dimensional mesh plot command in MATLAB is **mesh** **(X,Y,Z,C)**, where X, Y, and Z are matrices and C specifies the color of the grid lines. If **meshc(X,Y,Z,C)** is used, a contour plot is drawn beneath the mesh. If **meshz(X,Y,Z,C)** is used, a curtain plot is drawn beneath the mesh.

Consider the function $z = \cos(x)^*\sin(y)$ in the interval $-2 < x < 2$, $-2 < y < 2$. To draw a mesh or contour graph of this function, first form matrices x and y containing a grid of values in this range (see Fig. 2.11):

$\gg$ **dx = 1/3**

dx =
    0.3333

$\gg$ **dy = 1/3**

dy =
    0.3333

$\gg$ **dz = 1/3**

dz =
   0.3333

≫ **[x,y] = meshgrid(−2:dx:2, − 2:dy:2);**

We can evaluate this function at all the points in **x** and **y** with

≫ **z = cos(x) .* sin(y);**
≫ **mesh(z)**

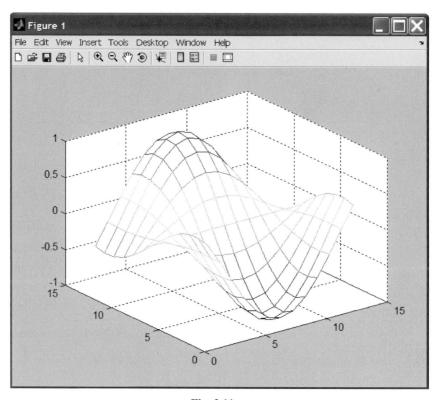

Fig. 2.11

The command **meshgrid(x,y)** transforms the domain specified by the vectors
x and **y** into arrays **X** and **Y** that can be used for the evaluation of functions of two
variables and three-dimensional mesh/surface plots. The rows of the output array
X are copies of the vector **x**, and the columns of the output array **Y** are copies of
the vector **y**. Note that the command **mesh(x,y,z)** would plot the actual values of
x and **y** used to generate **z**.

Another example of the use of the **meshgrid** command with the **mesh** family
of commands follows:

≫ **[x,y] = meshgrid(− 2:.2:2, − 2:.2:2);**
≫ **z = x .* exp(−x.^2 −y.^2);**

≫ **mesh(z)**
≫ **title('This is a 3-D plot of z = x * exp(−x^2 − y^2)')**

This family of commands is extremely useful for generating graphics such as carpet plots (see Fig. 2.12).

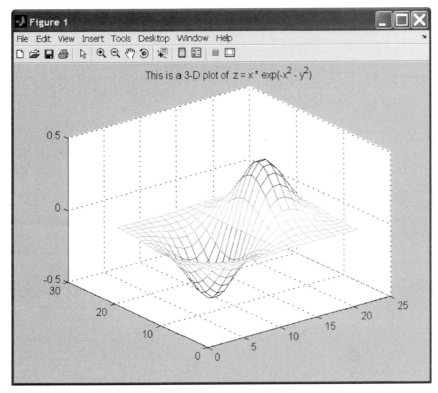

Fig. 2.12

2.8 Contour Diagrams

The command **contour(Z)** draws a contour plot of the matrix **Z**. The command **contourc(C)** calculates the contour matrix **Z** for use by the M-file **contour** to draw the actual contour plots. The command **contour3(Z)** produces a three-dimensional contour plot of a surface defined on a rectangular grid.

A quiver or needle plot of vectors with direction and magnitude are generated using the **quiver(X,Y,DX,DY)** command. This command draws arrows at every pair of elements in the matrices **X** and **Y**. The pairs of elements in matrices **DX** and **DY** determine the direction and relative magnitude of the arrows. A final trailing argument specifies line type and color using any legal line specification as described under the **plot** command.

As an example of **contour** and **quiver**, the function $z = \cos(x)^*\sin(y)$ will now be analyzed. The gradient of this function is easy to compute analytically:

$dz/dx = -\sin(x)^*\sin(y)$
$dz/dy = \cos(x)^*\cos(y)$

We can evaluate the partials for all points in **x** and **y** using the MATLAB expressions

>> **dx = 1/3; dy = 1/3;**
>> **[x,y] = meshgrid(−2:dx:2,−2:dy:2);**
>> **z = cos(x) .* sin(y);**
>> **dzx = −sin(x).*sin(y);**
>> **dzy = cos(x).*cos(y);**

If the gradient of a function is too complicated to compute analytically, or if we start with data arrays, the gradient can be computed numerically. The following is an example using the MATLAB function **gradient**:

[pzx,pzy] = gradient(z,dx,dy);

Overlaying a contour plot of the function and a quiver plot of the partials puts directional information on the contour plot. The command **hold on** retains the previous plot so that additional information can be overlaid.

Figure 2.13 shows the plot after the **contour(z)** command is used:

>> **contour(z)**

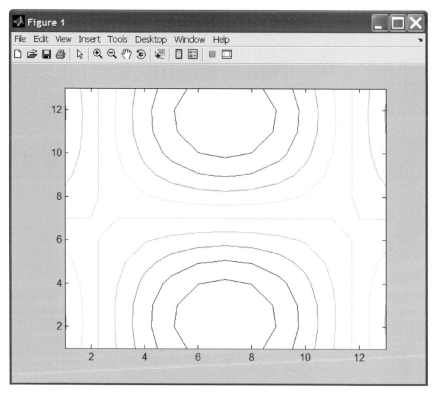

Fig. 2.13

Next the **quiver(dzx,dzy)** result is overlaid with **hold on** (see Fig. 2.14):

>> **contour(z), hold on**
>> **quiver(dzx,dzy), hold off**

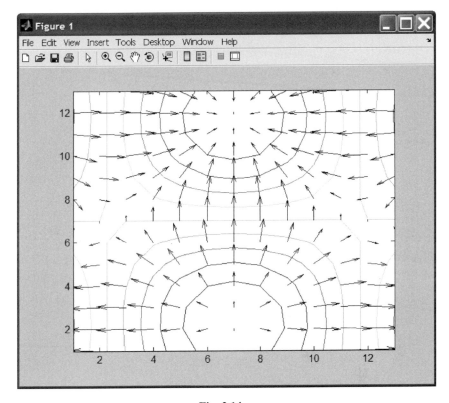

Fig. 2.14

Note that a comma allows multiple command lines to be placed on a single line:

>> **contour(z), hold on, quiver(pzx,pzy), hold off**

Both of these sets of commands give the same result.

2.9 Flow Diagrams

As an example of three-dimensional contour plotting using MATLAB, the **peaks** function is next plotted using the three-dimensional contour graphics command **contour3** along with the following (see Fig. 2.15):

>> **x = −3:0.125:3;**
>> **y = x;**

```
>> [X,Y] = meshgrid(x,y);
>> Z = peaks(X,Y);
>> contour3(X,Y,Z,20)
```

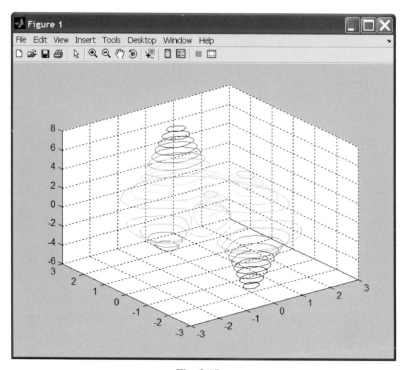

Fig. 2.15

2.10 Movies

The command **movie(M)** plays recorded movie frames. The command **movie(M,N)** plays the movie N times. If **N** is negative, it plays the movie once forward and once backward. The command **movie(M,N,FPS)** plays the movie at **FPS** frames per second. The default is 12 frames per second.

The command **moviein(N)** creates a matrix large enough to hold N frames for a movie.

The command **getframe** returns a column vector with one movie frame. The frame is a snapshot (pixmap) of the current axis.

The following example generates a movie with n frames:

```
>> z = peaks;
>> surf(z);
>> lim = axis;
>> M = moviein(20);
>> for j = 1:20, surf(sin(2*pi*j/20)*z,z), axis(lim), M(:,j) =getframe;,
end, movie(M,20)
```

Some of the resulting frames from this movie are shown in Figs 2.16 and 2.17. A MATLAB avi of this movie can also be embedded in electronic documents. You should also try these commands yourself to see the results of the **movie(m,n)** command in action!

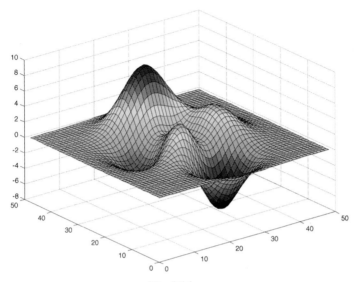

Fig. 2.16

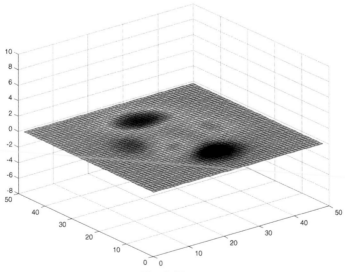

Fig. 2.17

Figure 2.18 is the movie saved using the **avifile** function.

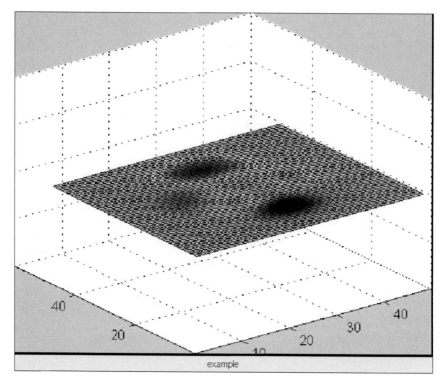

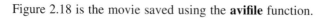

Fig. 2.18

This concludes the discussion on plotting and graphics.

2.11 Conclusion

This chapter introduced the reader to the plotting and graphics tools that are available in MATLAB as well as the methods for generating simple MATLAB animations.

Practice Exercises

2.1 Use MATLAB commands to generate the linear plot of $y = 5x^2$.
2.2 Use MATLAB commands to generate the semilogx plot of $y = 5x^2$.
2.3 Use MATLAB commands to generate the semilogy plot of $y = 5x^2$.
2.4 Use MATLAB commands to generate the log-log plot of $y = 5x^2$.

Generate the data using the commands

```
>> for ic=1:100;,x(ic)=(ic−1)/2;,y(ic)=5*x(ic)^2;end;
```

Notes

Introduction to MATLAB® Toolboxes

3.1 Introduction and Objectives

This chapter covers a few of the computational and graphics routines that are available in MATLAB® toolboxes. It emphasizes the analysis of aircraft dynamics and uses many commands from the Controls Toolbox.

Upon completion of this chapter, the reader will be able to identify and input some computational and graphics routines to conduct analysis of aircraft dynamics and input some commands from the Control Toolbox and related toolboxes.

3.2 Continuous Transfer Functions

Suppose we start with a plant description in transfer function form:

$$H(s) = \frac{0.2 \, s^2 + 0.3 \, s + 1}{(s^2 + .4 \, s + 1)(s + .5)}$$

We enter the numerator and denominator coefficients into MATLAB as vectors in descending powers of s:

> **num = [.2 .3 1];**
> **den1 = [1 .4 1];**
> **den2 = [1 .5];**

Remember that the Laplace notation is $dx/dt = \dot{X} = x^*s$.

3.2.1 Convolution

The denominator polynomial is the product of the two terms. We can use convolution to obtain the polynomial product:

> **den = conv(den1,den2)**
> den =
>   1.0000   0.9000   1.2000   0.5000

Working the problem out in longhand,

$$(s^2 + .4 \, s + 1)(s + .5) = s^3 + 0.9 \, s^2 + 1.2 \, s + 0.5$$

3.2.2 Print System

The transfer function can be printed in standard form using the **printsys** command. This is one of many commands added to MATLAB to provide some Mathematica-style symbolic math capabilities:

> **printsys(num,den)**

num/den =

$$\frac{0.2 \ s^2 + 0.3 \ s + 1}{s^3 + 0.9 \ s^2 + 1.2 \ s + 0.5}$$

Because s is the default symbol, the following gives the same result:

> **printsys(num,den,'s')**

The transfer function can also be printed in standard form using **tf**:

> **tf(num,den)**

Transfer function:

$$\frac{0.2 \ s^2 + 0.3 \ s + 1}{s^3 + 0.9 \ s^2 + 1.2 \ s + 0.5}$$

3.2.3 Damping

We can look at the natural frequencies and damping factors of the plant poles:

> **damp(den)**

Eigenvalue	Damping	Freq. (rad/s)
$-2.00\text{e-}001 + 9.80\text{e} - 001\text{i}$	$2.00\text{e} - 001$	$1.00\text{e} + 000$
$-2.00\text{e-}001 - 9.80\text{e} - 001\text{i}$	$2.00\text{e} - 001$	$1.00\text{e} + 000$
$-5.00\text{e} - 001$	$1.00\text{e} + 000$	$5.00\text{e} - 001$

Remember that the eigenvalues are roots of the denominator polynominal. Two of the roots have real coefficients only in their second-order representation. Here $i = sqrt(-1)$.

Other ways to generate this information will be demonstrated throughout this text.

3.2.4 Equivalent Continuous State-Space Model

A state-space representation can be obtained from a transfer function model by using the **tf2ss** command. The state-space representation is of the form

$$\dot{x} = Ax + Bu$$
$$y = Cx + Du$$

≫ **[a,b,c,d] = tf2ss(num,den)**

a =

$$\begin{matrix} -0.9000 & -1.2000 & -0.5000 \\ 1.0000 & 0 & 0 \\ 0 & 1.000 & 0 \end{matrix}$$

b =

1

0

0

c =

0.2000 0.3000 1.0000

d =

0

3.3 Root Locus

A root locus can be obtained by using the **rlocus** command:

≫ **rlocus(num,den);**

The same format rules apply for the time and frequency response functions shown in Figure 3.1.

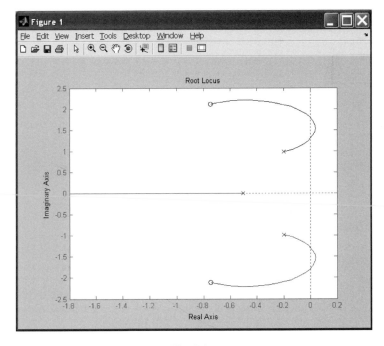

Fig. 3.1

Background materials on the equivalence between these plot types follow in Table 3.1 for twelve common transfer functions. These show the relationship between the open-loop transfer function, the Nyquist and Bode diagrams, the Nichols chart, and the root locus plot.

3.4 Step and Impulse Responses

For systems described in state space or by transfer functions, the step response is found by using the **step** command. Similarly, the impulse response is generated by using the **impulse** command (see Fig. 3.2):

≫ **step(num,den);**

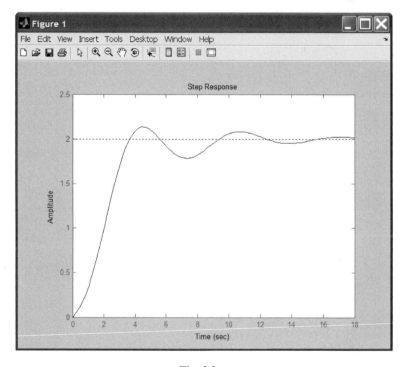

Fig. 3.2

The following commands would give the same results:

≫ **step(a,b,c,d,1);**   ≫ **step(sys);**

where the model **sys** is generated using either **sys = tf(num,den)** or **sys = ss(a,b,c,d)**. Also note that when using the state-space representation, the position of the input must be specified, or all inputs will be plotted.

Table 3.1 Comparison between Controls Analysis Methods

No.	Analysis method	When is this method used?	Important equations	Important variables	Typical plot	Remarks/summary of method
1	Time domain	When a specific time constant (T), setting time (ts), Max overshoot, and/or steady-state error (ess), ... are required.	1. Transfer function (TF) 2. Characteristic equation 3. Initial value theory 4. Final value theory 5. Laplace transform $$TF = \frac{\text{output}(s)}{\text{input}(s)}$$	T, ts Mp e_{ss}		Find the TF of the system, find its characteristic equation, find its roots, judge stability, find the damping ratio (ξ), the time constant (T), and the setting time (ts).
2	Root locus	When a specific damping ratio (ξ) and undamped natural frequency (ω_n) are required.	TF with gain (K) $$TF = \frac{G(s)}{1 + KG(s)H(s)}$$	K, ξ, ω_n		Find the closed-loop TF, vary the gain from 0 to infinity, plot the loci of roots in the s plane, then find the gain where the system becomes unstable. Then find the gain at a specific ξ and ω_n. Used to find closed-loop gain. An important tool in classical control system design.

Continued

Table 3.1   Comparison between Controls Analysis Methods (*Continued*)

No.	Analysis method	When is this method used?	Important equations	Important variables	Typical plot	Remarks/summary of method
3	Frequency response (e.g. Bode plot)	When a specific gain margin, phase margin, and/or bandwidth, ... are required.	$e_i = A_i \sin(\omega t)$ $e_o = A_o \sin(\omega t + \phi)$ $M = 20\log\dfrac{A_o}{A_i}\ (dB)$	M ϕ ω_b		Replace s with $j\omega$ in the TF, then find the magnitude and phase shift versus frequency. Used for both open-loop and closed-loop systems. Another important tool in classical control systems design.
4	State space	When the ODE is linear and the system is MIMO.	$\dot{X} = AX + BU$ $Y = CX + DU$	A, B, C, D, states	Same as 1 and 3	Used in most modern control system methods.

3.5   Bode Plot

The frequency response is found by using the Bode command (see Fig. 3.3):

≫ **bode(num,den);**   or   ≫ **bode(a,b,c,d,1);**

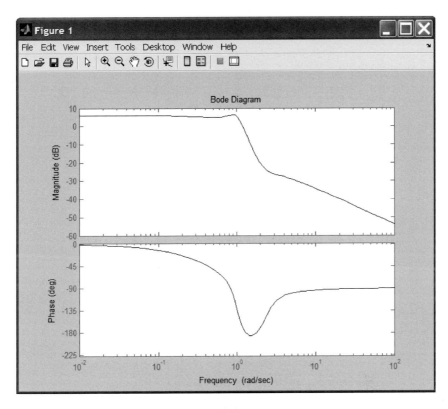

Fig. 3.3

To obtain a listing of the actual data values in the Bode plot, use the **bode** command:

≫ **[freq, amp, omega] = bode(a,b,c,d,1)**

The values in the vector amp are the magnitude of the output divided by the input magnitude at that frequency.

To generate the amplitude in decibels, use

≫ **ampdb = 20*log10(amp)**

Finally, user-specified omega values can be used as follows:

≫ **bode(a,b,c,d,1,omega);**   or   ≫ **bode(num,den,omega);**

3.6   Nichols Chart

The frequency response is also found by using the Nichols command (see Fig. 3.4):

≫ **nichols(a,b,c,d,1);**   or   ≫ **nichols(num,den);**

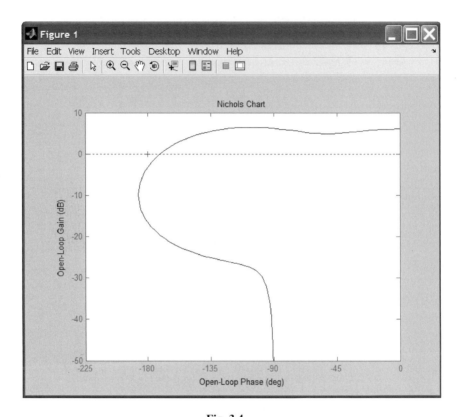

Fig. 3.4

This format is very good for graphically displaying the gain and phase margins.

3.7   Nyquist Chart

The frequency response plotted in the real-imaginary plane is found by using the Nyquist command (see Fig. 3.5):

≫ **nyquist(a,b,c,d,1);**   or   ≫ **nyquist(num,den);**

Here 90 degrees of phase lag is equivalent to $0 + 1i$ or the sqrt(-1). An out-of-phase output is equivalent to 180 degrees of phase lag or -1 on the Nyquist chart.

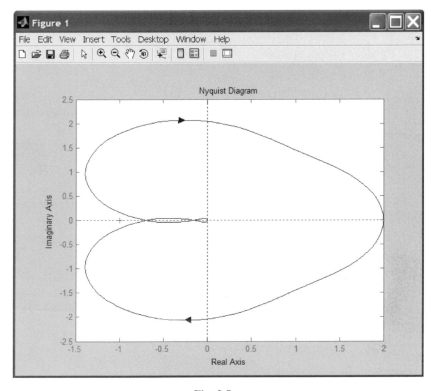

Fig. 3.5

3.8 Linear Quadratic Regulator

A linear quadratic regulator is now designed for this plant.
For control and state penalties,

```
≫ r = 1;
≫ q = eye(size(a))
q =
    1   0   0
    0   1   0
    0   0   1
```

The quadratic optimal gains, the associated Riccati solution, and the closed-loop
eigenvalues are

```
≫ [k,s,e] = lqr(a,b,q,r)

k =
    1.1983   1.2964   0.6180
```

s =

    1.1983   1.2964   0.6180
    1.2964   3.5400   1.8959
    0.6180   1.8959   2.1910

e =

    −0.7540
    −0.6721 + 1.0154i
    −0.6721 − 1.0154i

Other modern state-space control design methods are included in the Robust Control and μ-Synthesis Toolboxes.

3.9   State-Space Design

This file demonstrates MATLAB's ability in classical control system design by going through the design of a yaw damper for a jet transport aircraft using a state-space representation.

Define the jet transport model as Mach = 0.8 and h = 40,000 ft:

```
>> A = [−.0558    −.9968     .0802      .0415
          .598     −.115    −.0318      0
         −3.05      .388     −.4650     0
          0        0.0805    1          0];

>> B = [.0729      .0001
        −4.75      1.23
         1.53     10.63
         0         0];

>> C = [0 1 0 0
        0 0 0 1];

>> D = [0 0
        0 0];
>> states = 'beta yaw roll phi';
>> inputs = 'rudder aileron';
>> outputs = 'yaw-rate bank-angle';
>> printsys(A,B,C,D,inputs,outputs,states)
```

a =

	beta	yaw	roll	phi
beta	−0.05580	−0.99680	0.08020	0.04150
yaw	0.59800	−0.11500	−0.03180	0
roll	−3.05000	0.38800	−0.46500	0
phi	0	0.08050	1.00000	0

b =

	rudder	aileron
beta	0.07290	0.00010
yaw	−4.75000	1.23000
roll	1.53000	10.63000
phi	0	0

c =

	beta	yaw	roll	phi
yaw-rate	0	1.00000	0	0
bank-angle	0	0	0	1.00000

d =

	rudder	aileron
yaw-rate	0	0
bank-angle	0	0

These are the state-space matrices for a jet transport during cruise flight. The model has two inputs and two outputs. The units are radians for beta (sideslip angle) and phi (bank angle) and radians per second for yaw (yaw rate) and roll (roll rate). The rudder and aileron deflections are in degrees.

This model has one set of eigenvalues that are lightly damped. They correspond to what is called the Dutch roll mode. We need to design a compensator that increases the damping of these poles:

≫ **disp('Open Loop Eigenvalues'), damp(A);**

Open-Loop Eigenvalues

Eigenvalue	Damping	Freq. (rad/s)
−0.0073	1.0000	0.0073
−0.0329 + 0.9467i	0.0348	0.9472
−0.0329 − 0.9467i	0.0348	0.9472
−0.5627	1.0000	0.5627

Our design criteria are to provide damping ratio $\zeta > 0.35$ with natural frequency $\omega_n < 1.0$ radian/second. We want to design the compensator using classical methods.

Let us do some open-loop analysis to determine possible control strategies. Time response (we could use **step** or **impulse** here) (see Fig. 3.6):

≫ **step(A,B,C,D);**

Note that all inputs and outputs were plotted using this command (see Fig. 3.7).

≫ **impulse(A,B,C,D);**

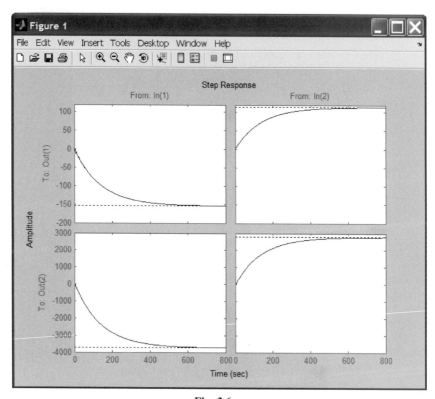

Fig. 3.6

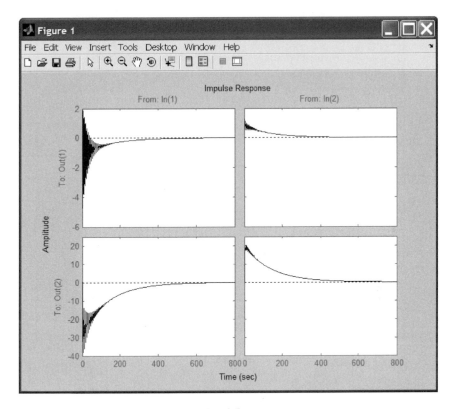

Fig. 3.7

The time responses show that the system is indeed lightly damped. But the time frame is much too long. Let us look at the response over a smaller time frame. Define the time vector T before invoking **impulse.**

Define a time vector from 0 to 20 seconds in steps of 0.2:

$\gg$ **T = 0:0.2:20;**

Plot the responses as separate graphs (see Fig. 3.8):

```
>> subplot(221), impulse(A,B,C(1,:),D(1,:),1,T);
>> title('Input 1 Output 1')
>> subplot(222), impulse(A,B,C(2,:),D(2,:),1,T);
>> title('Input 1 Output 2')
>> subplot(223), impulse(A,B,C(1,:),D(1,:),2,T);
>> title('Input 2 Output 1')
>> subplot(224), impulse(A,B,C(2,:),D(2,:),2,T);
>> title('Input 2 Output 2')
>> subplot
```

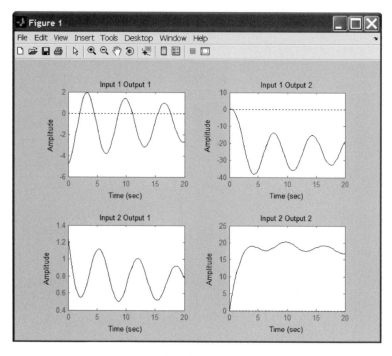

Fig. 3.8

Look at the plot from aileron (input 2) to bank angle (output 2). The aircraft is oscillating around a nonzero bank angle. The aircraft is turning in response to an aileron impulse. This behavior will be important later.

Typically yaw dampers are designed using yaw rate as the sensed output and rudder as the input. Let us look at that frequency response (see Fig. 3.9):

> **bode(A,B,C(1,:),D(1,:),1);**

From this frequency response we see that the rudder is effective around the lightly damped Dutch roll mode (at 1 radian/second).

To make the design easier, extract the subsystem from rudder to yaw rate. Extracting the subsystem with input 1 and output 1,

> **[a,b,c,d] = ssselect(A,B,C,D,1,1);**

Let us do some designs. The simplest compensator is a gain. We can determine values for this gain using the root locus (see Fig. 3.10):

> **rlocus(a,b,c,d);**

Oops. Looks like we need positive feedback (negative feedback is assumed) (see Fig. 3.11):

> **rlocus(a,b,−c,−d); sgrid**

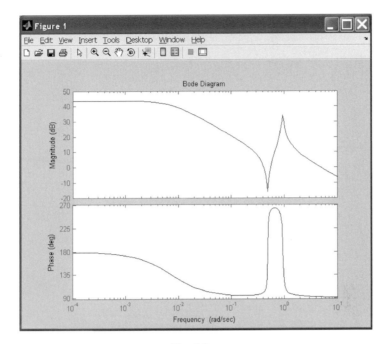

Fig. 3.9

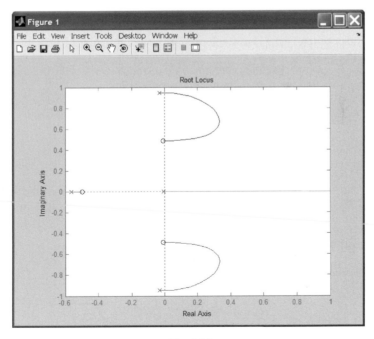

Fig. 3.10

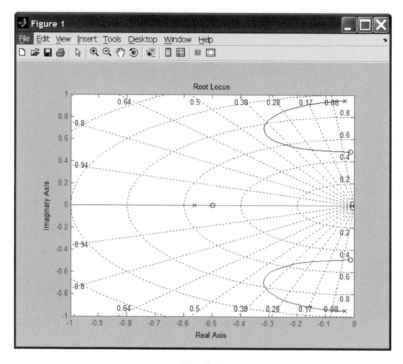

Fig. 3.11

That looks better. Using just simple feedback, we can achieve a damping ratio of $\zeta = 0.45$.

Now it is your turn. Use the capability within **rlocus** to select the point on the root locus providing the maximum closed-loop damping.

Select a point in the graphics window using the mouse. MATLAB will return with the following (see Fig. 3.12):

selected point =

  $-0.3114 + 0.6292i$

Note that multiple points may be selected in this manner from the figure window. We will next use MATLAB to convert this result into a text string to display with the closed-loop system gains:

≫ **disp(['You chose gain: ',num2str(k)]),damp(esort(poles));**

You chose gain 0.26798.

Eigenvalue	Damping	Freq. (rad/s)
$-3.08e-001 + 6.30e-001i$	$4.39e-001$	$7.02e-001$
$-3.08e-001 - 6.30e-001i$	$4.39e-001$	$7.02e-001$
$-3.25e-001$	$1.00e+000$	$3.25e-001$
$-9.67e-001$	$1.00e+000$	$9.67e-001$

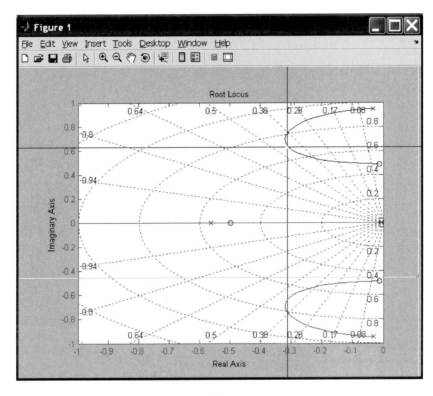

Fig. 3.12

Let us form the closed-loop system so that we can analyze the design:

```
>> [ac,bc,cc,dc] = feedback(a,b,c,d,[ ],[ ],[ ],−k);
ac =
    −0.0558    −0.9773     0.0802    0.0415
     0.5980    −1.3879    −0.0318    0
    −3.0500     0.7980    −0.4650    0
          0     0.0805     1.0000    0
bc =
     0.0729
    −4.7500
     1.5300
          0
cc =
     0    1    0    0
dc =
          0
```

These eigenvalues should match the ones you chose (see Fig. 3.13):

```
>> disp('Closed loop eigenvalues'), damp(ac);
>> impulse(ac,bc,cc,dc,1,T);
```

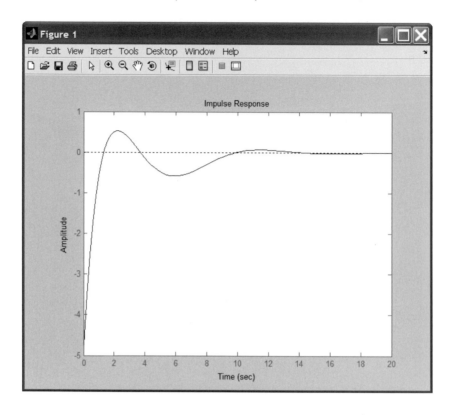

Fig. 3.13

This response looks pretty good. Let us close the loop on the original model and see how the response from the aileron looks. This is accomplished using feedback of output 1 to input 1 of plant.

Feedback with selection vectors assumes positive feedback (see Fig. 3.14):

```
≫ [Ac,Bc,Cc,Dc] = feedback(A,B,C,D,[ ],[ ],[ ],k,[1],[1]);
≫ disp('Closed loop eigenvalues'), damp(Ac);
≫ T = 0:0.2:20;
≫ subplot(221), impulse(Ac,Bc,Cc(1,:),Dc(1,:),1,T);
≫ title('Input 1 Output 1')
≫ subplot(222), impulse(Ac,Bc,Cc(2,:),Dc(2,:),1,T);
≫ title('Input 1 Output 2')
≫ subplot(223), impulse(Ac,Bc,Cc(1,:),Dc(1,:),2,T);
≫ title('Input 2 Output 1')
≫ subplot(224), impulse(Ac,Bc,Cc(2,:),Dc(2,:),2,T);
≫ title('Input 2 Output 2')
≫ subplot
```

Look at the plot from aileron (input 2) to bank angle (output 2). When we move the aileron, the system no longer continues to bank like a normal aircraft.

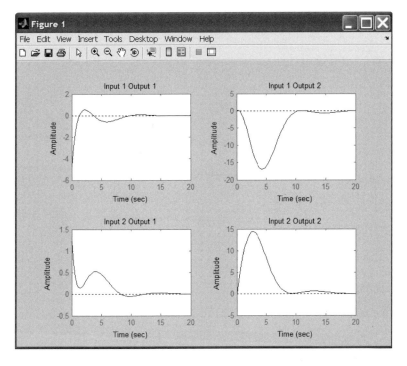

Fig. 3.14

We have overstabilized the spiral mode. The spiral mode is typically a very slow mode and allows the aircraft to bank and turn without constant aileron input. Pilots are used to this behavior and will not like our design. Our design has moved the spiral mode so that it has a faster frequency.

What we need to do is make sure the spiral mode doesn't move farther into the left half-plane when we close the loop. One way to fix this problem is to use a washout filter, i.e.,

$$H(s) = \frac{Ks}{(s + a)}$$

Choosing **a = 0.333** for a time constant of 3 seconds, form the washout:

```
≫ [aw,bw,cw,dw] = zp2ss([0],[−.333],1)
```

aw =
  −0.3330

bw =
  1

cw =
  −0.3330

dw =
  1

Connect the washout in series with our design model:

≫ **[a,b,c,d] = series(a,b,c,d,aw,bw,cw,dw)**

a =

−0.3330	0	1.0000	0	0
0	−0.0558	−0.9968	0.0802	0.0415
0	0.5980	−0.1150	−0.0318	0
0	−3.0500	0.3880	−0.4650	0
0	0	0.0805	1.0000	0

b =

```
      0
 0.0729
-4.7500
 1.5300
      0
```

c =

−0.3330	0	1.0000	0	0

d =

```
0
```

Do another root locus (see Fig. 3.15):

≫ **rlocus(a,b,−c,−d); sgrid**

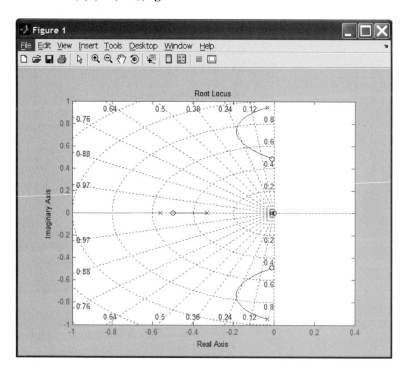

Fig. 3.15

Now the maximum damping is $\zeta = 0.25$.

Now use **rlocus** to choose the gain for maximum damping. Selecting a point in the graphics window gives the following result (see Fig. 3.16):

selected_point =

  $-0.1818 + 0.6956i$

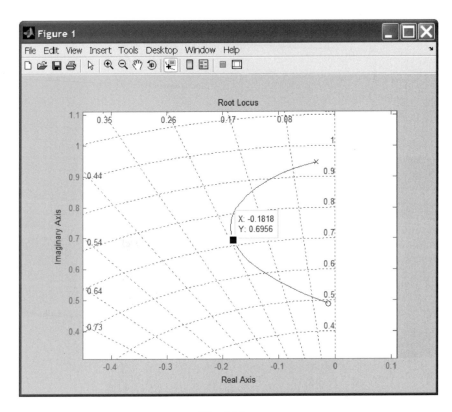

Fig. 3.16

≫ **disp(['You choose gain: ',num2str(k)]), damp(esort(poles));**

You choose gain 0.23263 (see Fig. 3.17).

Eigenvalue	Damping	Freq. (rad/s)
$-4.07e - 003$	$1.00e + 000$	$4.07e - 003$
$-1.82e - 001 + 6.96e - 001i$	$2.53e - 001$	$7.19e - 001$
$-1.82e - 001 - 6.96e-001i$	$2.53e - 001$	$7.19e - 001$
$-4.71e - 001$	$1.00e + 000$	$4.71e - 001$
$-1.24e + 000$	$1.00e + 000$	$1.24e + 000$

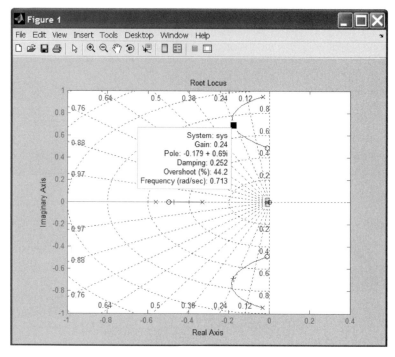

Fig. 3.18

Look at the closed-loop response (see Fig. 3.18):

```
≫ [ac,bc,cc,dc] = feedback(a,b,c,d,[ ],[ ],[ ],−k)
≫ impulse(ac,bc,cc,dc,1,T);
```

ac =
```
   -0.3330        0        1.0000         0         0
   -0.0056   -0.0558       -0.9798    0.0802    0.0415
    0.3680    0.5980       -1.2200   -0.0318         0
   -0.1185   -3.0500        0.7439   -0.4650         0
         0         0        0.0805    1.0000         0
```

bc =
```
         0
    0.0729
   -4.7500
    1.5300
         0
```

cc =
```
   -0.3330        0        1.0000         0         0
```

dc =
```
         0
```

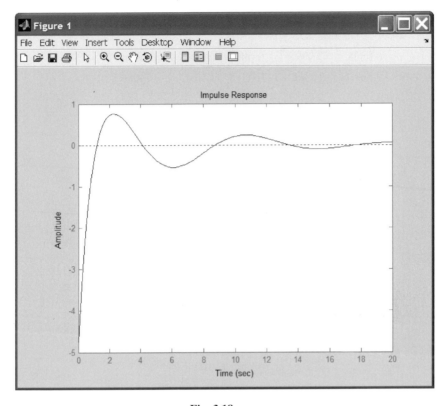

Fig. 3.18

Now form the controller (washout + gain):

≫ **[aw,bw,cw,dw] = series(aw,bw,cw,dw,[ ],[ ],[ ],k)**

    aw =
      −0.3330

    bw =
       1

    cw =
      −0.0775

    dw =
       0.2326

Close the loop with the original model:

≫ **[Ac,Bc,Cc,Dc]= feedback(A,B,C,D,aw,bw,cw,dw,[1],[1])**

Ac =

−0.0558	−0.9798	0.0802	0.0415	−0.0056
0.5980	−1.2200	−0.0318	0	0.3680
−3.0500	0.7439	−0.4650	0	−0.1185
0	0.0805	1.0000	0	0
0	1.0000	0	0	−0.3330

Bc =

0.0729	0.0001
−4.7500	1.2300
1.5300	10.6300
0	0
0	0

Cc =

0	1	0	0	0
0	0	0	1	0

Dc =

0	0
0	0

The final closed-loop time responses are plotted:

≫ **subplot(221), impulse(Ac,Bc,Cc(1,:),Dc(1,:),1,T);**
≫ **title('Input 1 Output 1')**
≫ **subplot(222), impulse(Ac,Bc,Cc(2,:),Dc(2,:),1,T);**
≫ **title('Input 1 Output 2')**
≫ **subplot(223), impulse(Ac,Bc,Cc(1,:),Dc(1,:),2,T);**
≫ **title('Input 2 Output 1')**
≫ **subplot(224), impulse(Ac,Bc,Cc(2,:),Dc(2,:),2,T);**
≫ **title('Input 2 Output 2')**

Although we didn't quite meet the criteria, our design increased the damping of the system substantially (see Fig. 3.19).

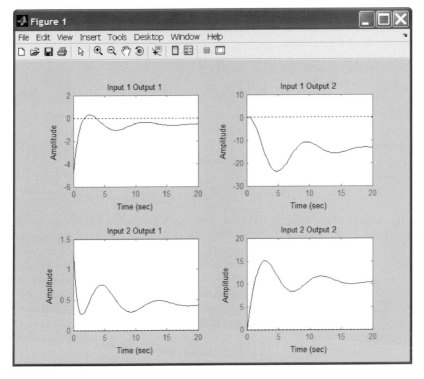

Fig. 3.19

3.10   Digital Design

This example demonstrates MATLAB's ability in digital system design by synthesizing a computer hard disk read/write head position controller.

Using Newton's law, we can model the simplest model for the read/write head with the following differential equation:

$$I^*\text{theta\_ddot} + C^*\text{theta\_dot} + K^*\text{theta} = Ki^*i$$

where I is the inertia of the head assembly; C is the viscous damping coefficient of the bearings; K is the return spring constant; Ki is the motor torque constant; theta_ddot, theta_dot, and theta are the angular acceleration, angular velocity, and position of the head; and i is the input current.

Taking the Laplace transform, the transfer function is

$$H(s) = \frac{K\,i}{I\,s^2 + C\,s + K}$$

Using the values $I = 0.01$ Kg m^2, $C = 0.004$ Nm/(radians/second), $K = 10$ Nm/radians, and $Ki = 0.05$ Nm/radians, form the transfer function description of this system:

```
≫ I = .01; C = 0.004; K = 10; Ki = .05;
≫ NUM = [Ki];
≫ DEN = [I C K];
≫ printsys(NUM,DEN,'s');
```

num/den =

$$\frac{0.05}{0.01 \ s^2 + 0.004s + 10}$$

Our task is to design a digital controller that can be used to provide accurate positioning of the read/write head. We will do the design in the digital domain.

First we must discretize our plant because it is continuous. MATLAB has several methods available for this discretization using the function **c2dm.** Let us compare all the methods and choose the best one. Note that starting in version 6.1 the continuous system is represented using the new sys format, whereas the discrete systems still use the older transformation format. Because of the number of toolboxes and commands in MATLAB, these types of inconsistencies can be encountered.

Use the sample time $\tau_s = 0.005$ (5 mseconds) (see Fig. 3.20):

```
≫ Ts = 0.005;
≫ sys = tf(NUM,DEN);
≫ clf;
≫ [magar, phasear,w] = bode(sys);
≫ bode(sys);
≫ title('Bode of Continuous Open Loop Hard Disk Drive');
```

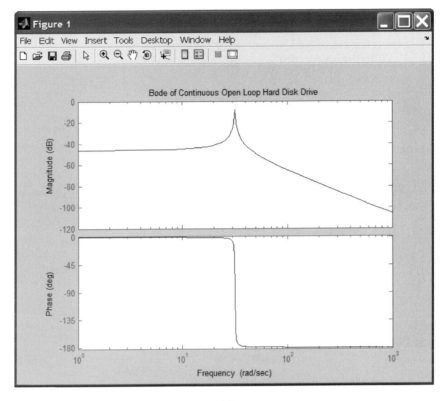

Fig. 3.20

Because of the different model structures, the new array format is converted into the older vector format for comparison with the discrete system responses:

```
≫ ender = size(magar);
≫ for i = 1:ender(3)
≫    mag(i) = magar(1,1,i);
≫    phase(i) = phasear(1,1,i);
≫ end
```

Now plot the results as a comparison (see Fig. 3.21):

```
≫ [num,den] = c2dm(NUM,DEN,Ts,'zoh');
≫ [mzoh,pzoh] = dbode(num,den,Ts,w);
≫ [num,den] = c2dm(NUM,DEN,Ts,'foh');
≫ [mfoh,pfoh] = dbode(num,den,Ts,w);
≫ [num,den] = c2dm(NUM,DEN,Ts,'tustin');
≫ [mtus,ptus] = dbode(num,den,Ts,w);
≫ [num,den] = c2dm(NUM,DEN,Ts,'prewarp',30);
≫ [mpre,ppre] = dbode(num,den,Ts,w);
≫ [num,den] = c2dm(NUM,DEN,Ts,'matched');
≫ [mmat,pmat] = dbode(num,den,Ts,w);
```

```
>> subplot(211);
>> semilogx(w,20*log10(mag)), hold on;
>> semilogx(w,20*log10(mzoh),w,20*log10(mfoh),w,
      20*log10(mtus),w,20*log10(mpre),w,20*log10(mmat))
```

```
>> hold off;
>> xlabel('Frequency (rad/sec)'), ylabel('Gain dB');
>> title('c2d Comparison Plot');
```

```
>> subplot(212);
>> semilogx(w,phase), hold on;
>> semilogx(w,pzoh,w,pfoh,w,ptus,w,ppre,w,pmat), hold off;
>> xlabel('Frequency (rad/sec)'), ylabel('Phase deg');
```

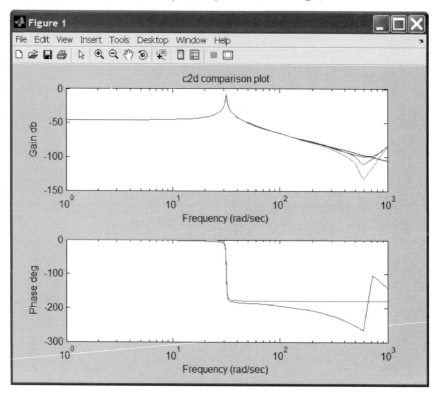

Fig. 3.21

Looking at these plots, we see that they all seem to be pretty good matches to the continuous response. However, the matched pole-zero method gives a marginally better match to the continuous response.

Note that Franklin and Powell's book *Digital Control Systems* provides an excellent discussion of these methods.

Now analyze the discrete system. A summary of some of the continuous-to-discrete mappings is given in Table 3.2.

Table 3.2   Discrete representations of H(s) = a/(s + 2)

Forward rectangular rule	$H(s) = \dfrac{a}{(z-1)/T + a}$
Backward rectangular rule	$H(s) = \dfrac{a}{(z-1)/Tz + a}$
Trapezoid rule or Tustin's bilinear rule	$H(s) = \dfrac{a}{(2/T)[(z-1)/(z+1) + a]}$

A brief summary of Table 3.2 is that an approximation to the frequency variable can be used to provide a mapping from a continuous transfer function to a discrete transfer function or from the s plane to the z plane. These approximations are given in Table 3.3 for each of the preceding methods.

Additional insights into these mappings may be gained by considering them graphically. In the continuous plane, the $s = j\omega$-axis is the boundary separating the poles of stable systems from the poles of unstable systems. We will next look at the three mapping rules and will examine how the stable left-half system appears in the z plane. The inverse mappings from the previous table are given in Table 3.4.

By using the variable substitution $s = j\omega$ in the preceding equations, the boundaries to the shaded z-plane regions are generated for each of these three cases. These maps are given in Figs 3.22–3.24.

Another very effective method for obtaining the discrete time equivalent to a continuous transfer function is a pole-zero mapping. If we take the z transforms of the samples of a continuous signal e(t), then the poles of the discrete transform are related to the poles of the continuous transform by $z = e^{sT}$. The idea of the

Table 3.3   Approximate continuous to discrete mappings

Forward rectangular rule	$s \sim \dfrac{z-1}{T}$
Backward rectangular rule	$s \sim \dfrac{z-1}{Tz}$
Trapezoid rule or Tustin's bilinear rule	$s \sim \dfrac{2(z-1)}{T(z+1)}$

Table 3.4   Approximate discrete to continuous mappings

Forward rectangular rule	$z \sim 1 + Ts$
Backward rectangular rule	$z \sim \dfrac{1}{1 - Ts}$
Trapezoid rule or Tustin's bilinear rule	$z \sim \dfrac{1 + Ts/2}{1 - Ts/2}$

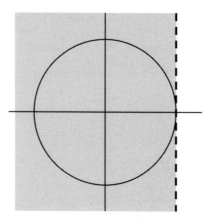

Fig. 3.22   Forward Rectangular Rule.

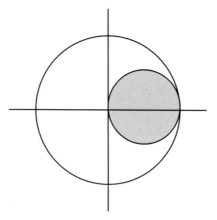

Fig. 3.23   Backward Rectangular Rule.

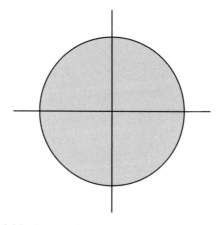

Fig. 3.24   Trapezoid Rule or Tustin's Bilinear Rule.

pole-zero mapping technique is that this same relation can also be applied to the system's zeros. The following are used to apply this technique.

1) All poles of H(s) are mapped by $z = e^{sT}$.

2) All finite zeros of H(s) are also mapped by $z = e^{sT}$.

3) All zeros of H(s) at $s = \infty$ are mapped to the point $z = -1$.

4) If a unit delay in the digital system's response is required, one zero of H(s) at $s = \infty$ is mapped into $z = \infty$.

5) The gain of the digital system is selected to match that of the continuous system at either the band center or at a critical frequency. Often this critical frequency is at $s = 0$.

Application of the preceding rules to

$$H(s) = \frac{a}{s + a}$$

results in the discrete transfer function:

$$H(z) = \frac{1 - e^{-aT}}{z - e^{-aT}}$$

A final discretization technique is the hold equivalence. The purpose of samplers is to work on only the system's outputs during each discrete time interval. Thus, the input-output behavior can be realized as a transfer function. The discrete equivalent is generated by first approximating e(t) from the samples e(k) and then running this time sequence through H(s). MATLAB contains techniques for taking this sequence and holding them through each discrete time interval to produce a continuous signal.

Now we will analyze one of the discrete system equivalents.

Displaying the discrete system generated using the matched technique,

≫ **printsys(num,den,'z')**

num/den =

$$\frac{6.2308e - 005\ z + 6.2308e - 005}{z^2 - 1.9731\ z + 0.998}$$

Plot the step response:

≫ **dstep(num,den);**

The system oscillates quite a bit. This is probably due to very light damping. We can check this by computing the open-loop eigenvalues (see Fig. 3.25).

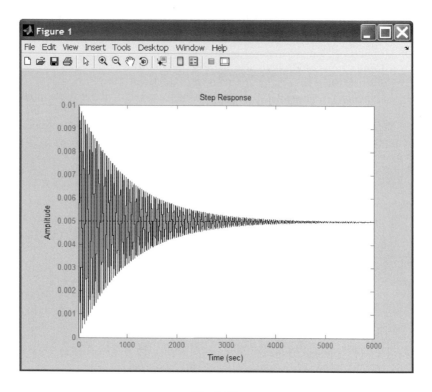

Fig. 3.25

>> disp('Open loop discrete eigenvalues'), ddamp(den,Ts);

Open-loop discrete eigenvalues

Eigenvalue	Magnitude	Equiv. Damping	Equiv. Freq. (rad/s)
0.9865 + 0.1573i	0.9990	0.0063	31.6228
0.9865 − 0.1573i	0.9990	0.0063	31.6228

This concludes our chapter on the use of toolboxes, detailing their use in dynamic systems analysis.

3.11 Conclusion

This chapter introduced the reader to a few of the computational and graphics routines available in MATLAB toolboxes. Many other MATLAB toolboxes are also available to the reader.

Practice Exercises

Enter the transfer function

$$H(s) = \frac{(s+3)}{(s^2 + 4s - 12)}$$

3.1 Calculate the partial fraction expansion using **residue**.
3.2 Convert to state space.
3.3 Use **c2d** to generate the discrete-time equations, with T = 0.5 seconds.
3.4 Use **tf2zp** to convert to a zero-pole-gain transfer function.
3.5 Convert the zero-pole-gain representation into state space.
3.6 Generate a Bode plot.
3.7 Generate a root locus.
3.8 Generate a Nyquist plot.

Notes

4
Introduction to MATLAB® Cells, Structures, and M-Files

4.1 Introduction and Objectives

This chapter covers some of the other data types that are available in MATLAB®. It also covers conditional statements and programming M-files.

Upon completion of this chapter, the reader will be able to identify other data types available in MATLAB (cells and structures) and develop both script and function M-files using conditional statements.

4.2 Cells

Cell arrays are data types that allow a user to store arbitrary data within each of the values of its array. For example, the following is a valid cell array:

```
>> myCell{[1 2 3;4 5 6],'Text in second index','embedded cell'}}

myCell =
    [2×3 double]   [1×24 char]   {1×1 cell}
```

Notice that any data type can be located in any index of the array. Cells are most useful as a place to store strings that have different lengths. To view each cell type the following command:

```
>> myCell{1}

ans =
    1   2   3
    4   5   6

>> myCell{2}
ans =
Text in the second index
```

95

≫ **myCell{3}**

ans =

   'embedded cell'

To view the second row of the array stored in cell 1, type the following command:

≫ **myCell{1}(2, :)**

ans =

   4 5 6

Two useful commands to examine the structure of cells and the data stored in the cells are **cellplot** and **celldisp** (see Fig. 4.1).

To examine the structure of **myCell**, use the command:

≫ **cellplot(myCell)**

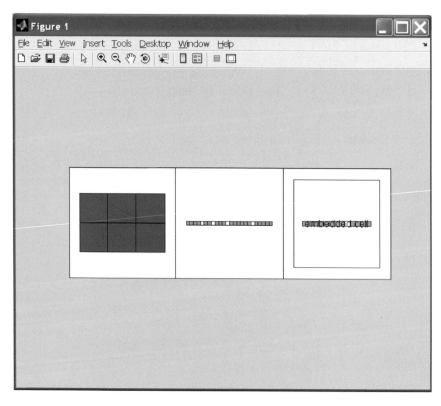

Fig. 4.1

To examine the data stored within **myCell**, use the command

≫ **celldisp(myCell)**

myCell{1} =

    1   2   3
    4   5   6

myCell{2} =

Text in the second index

myCell{3}{1} =

embedded cell

This concludes the section on cells.

4.3 Structures

Structures are data types that allow a user to store data in a field/value methodology. They are similar to records in a database. The following is an example of a structure that stores data in useful groupings:

≫ **Data.X.Name='AOA';**
≫ **Data.X.Units='Deg';**
≫ **Data.X.Values=1:10;**
≫ **Data.Y.Name='Cl';**
≫ **Data.Y.Units='';**
≫ **Data.Y.Values=[11:20]/10;**

≫ **Data.X**

ans =

    Name: 'AOA'
    Units: 'Deg'
    Values : [1 2 3 4 5 6 7 8 9 10]

≫ **Data.Y**

ans =

    Name: 'Cl'
    Units: ''
    Values : [1.1 1.2 1.3 1.4 1.5 1.6 1.7 1.8 1.9 2]

Structures really become useful when you have arrays of them. You can create additional elements of the **Data** array to hold other sets of data:

≫ **clear**
≫ **Data(2).X.Name='AOA';**
≫ **Data(2).X.Units='Deg';**

```
≫ Data(2).X.Values=1:10;
≫ Data(2).Y.Name='Cd';
≫ Data(2).Y.Units='';
≫ Data(2).Y.Values=[1:10]/1000;
≫ Data
```

Data =

1 × 2 struct array with fields:

   X

   Y

Data now has two elements in it containing data for **Cl** versus **AOA** and **Cd** versus **AOA**. This structure could be increased in dimension for additional forces and moments.

4.4 M-Files

MATLAB provides a full programming language that enables you to write a series of MATLAB statements into a file and then execute them with a single command. You write your program in an ordinary text file, giving the file a name of **filename.m**. The term you use for filename becomes the new command that MATLAB associates with the program. The file extension of **.m** makes this a MATLAB M-file. M-files can be scripts that simply execute a series of MATLAB statements, or they can be functions that also accept arguments and produce output. You create M-files using a text editor and then use them as you would any other MATLAB function or command.

There are two kinds of M-files: 1) script M-files and 2) function M-files. Script M-files do not accept input arguments or return output arguments and operate on data in the workspace. Scripts are the simplest kind of M-file because they have no input or output arguments. Function M-files accept input arguments and return output arguments and use internal variables (local to the function) by default.

Let us construct a MATLAB program or script M-file that generates the movie from Sec. 2.10 of Chapter 2. We will also examine the commands **if** and **for**, which are often used in M-files.

M-files are ordinary text files that you create using a text editor. MATLAB provides a built-in editor, although you can use any text editor you like. An advantage to using the build-in editor is that it can help you execute and debug your M-file.

Use the pull-down menu **File** to create a new M-file. Select **File**, then select **New**, and finally select **M-file** by clicking on the right button. This opens the MATLAB file editor. After entering the program (the five MATLAB commands given at the end of Chapter 2 of this book), select **File** and then the **Save As** option. Enter the filename (for example, **RDCMovie.m**) into the upper left field of the **Save As** window. This name must end in .m so that MATLAB recognizes it as an M-file.

To run the M-file RDCMovie.m in MATLAB, type

   ≫ **RDCMovie**

To modify an existing M-file, select **File** and **Open M-file** and then choose the desired M-file.

M-files can also be created using a normal text editor. The filename must end in .m for MATLAB to recognize the file as an M-file. The .m portion of the filename is not required when executing the file within the MATLAB **Command Window**. Also note that the .m portion of the filename is not displayed when using the **what** command.

Note that a % is used for comments. Everything after the % on that line is ignored during execution.

Continuations of lines are denoted with three periods .... For example, the vector **G** can be entered using the following command, which includes a continuation:

> G=[1.11 2.22 3.33 4.44 5.55 6.66 7.77 ... 8.88 9.99];

4.4.1 *If*

The **if** command conditionally executes statements. The simple form is

if **variable, statements** end

The statements are executed if the variable has all nonzero elements. The variable is usually the result of

'**expr' rop 'expr'**

where **rop** is ==, <, >, <=, >=, or ~=.
Other operators are

& = and
| = or
~ = not

For example, I and J must either be real positive integers or logical:

if I=J
  A(I,J)=2;
elseif abs(I − J)=1
  A(I,J)=−1;
else
  A(I,J)=0;
end
A

4.4.2 *For*

The **for** command repeats statements a specific number of times. An example of this is

M=4;
N=5;
for I=1:M

```
        for J=1:N
            A(I,J)=1/(I + J − 1);
        end
    end
    A
```

```
A =
    1.0000   0.5000   0.3333   0.2500   0.2000
    0.5000   0.3333   0.2500   0.2000   0.1667
    0.3333   0.2500   0.2000   0.1667   0.1429
    0.2500   0.2000   0.1667   0.1429   0.1250
```

4.4.3 While

The **while** command repeats statements an indefinite number of times. Here is an example that computes the first integer n for which n! (that is n factorial) is a 10 digit number:

```
n = 1;
    while prod(1:n) < 1.e10, n=n + 1;, end
n
```

This produces the result n = 14.

4.4.4 Strings

Text strings are entered into MATLAB surrounded by single quotes. For example,

```
≫ s = 'Hello'
```

results in

```
s =
    Hello
```

4.4.5 Function M-Files

This section shows you the basic parts of a function M-file, so that you can familiarize yourself with MATLAB programming and get started with some examples:

```
function f = fact(n) % Function definition line
% FACT Factorial. % H1 line
% FACT(N) returns the factorial of N, H! % Help text
% usually denoted by N!
% Put simply, FACT(N) is PROD(1:N).

f = prod(1:n); % Function body
```

This function has some elements that are common to all MATLAB functions.

1) A function definition line: This line defines the function name and the number and order of input and output arguments.

2) An H1 line: H1 stands for help 1 line. MATLAB displays the H1 line for a function when you use look for or request help on an entire directory.

3) Help text: MATLAB displays the help text entry together with the H1 line when you request help on a specific function.

4) The function body: This part of the function contains code that performs the actual computations and assigns values to any output arguments.

You can provide user information for the programs you write by including a help text section at the beginning of your M-file. This section starts on the line following the function definition and ends at the first blank line. Each line of the help text must begin with a comment character %. MATLAB displays this information whenever you type **help m-file_name**.

You can also make help entries for an entire directory by creating a file with the special name **Contents.m** that resides in the directory. This file must contain only comment lines; that is, every line must begin with a percent sign. MATLAB displays the lines in a **Contents.m** file whenever you type help **directory_name**.

If a directory does not contain a **Contents.m** file, typing **help directory_name** displays the first help line (the H1 line) for each M-file in the directory.

M-files are ordinary text files that you create using a text editor. MATLAB provides a built-in editor, although you can use any text editor you like. To open the editor on a PC, from the **File** menu, choose **New** and then **M-File**. Another way to edit an M-file is from the MATLAB command line using the **edit** function. For example, **edit my_mfile** opens the editor on the file **my_mfile.m**. Omitting a filename opens the editor on an untitled file.

As an example, let us construct an M-file function that accepts a structure containing data and then generates a plot using this data.

First we create the function to do the plotting as follows:

```
function []=myplot(DataArray);
% []=myplot(DataArray)
% Plots data contained in the structure 'DataArray'
%
plot(DataArray.X.Values,DataArray.Y.Values)
xText=[DataArray.X.Name ' (' DataArray.X.Units ')'];
xlabel(xText);
yText=[DataArray.Y.Name ' (' DataArray.Y.Units ')'];
ylabel(yText);
```

Next we create the data as in Sec. 4.3:

```
Data.X.Name='AOA';
Data.X.Units='Deg';
Data.X.Values=1:10;
Data.Y.Name='Cl';
Data.Y.Units='';
Data.Y.Values=[11:20]/10;
```

Finally we call the new function (see Fig. 4.2)

myplot(Data(1));

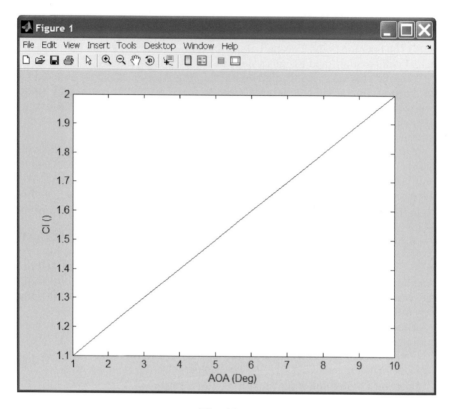

Fig. 4.2

To see commands as they are executing, use the **echo** command. The command **echo off** turns off this feature.

This concludes the discussion on M-files.

4.5 Conclusion

This chapter introduced the reader to MATLAB cells and structures. It also covered conditional statements and programming in MATLAB using M-files.

Practice Exercises

4.1   Construct a 2-by-2 cell array **A** as a database matrix containing the following information:

Location: Edwards Air Force Base
Date: 13 June 2006
Times: 6 a.m., 9 a.m., and 12 noon
Temperatures: 76, 79, and 78°F at 6 a.m.
Temperatures: 86, 92, and 89°F at 9 a.m.
Temperatures: 97, 102, and 97°F at 12 noon

4.2   Construct a structure array database containing the following information:

a)   Name
b)   Employee number
c)   E-mail address
d)   The vector [1 2 3 4 5]

4.3   Generate an M-file to calculate a rocket's trajectory and then save the data in the array **data**. The first column is the time, and the second column is the rocket height. In addition, generate an X-Y plot of the rocket's trajectory. Complete the following computations using MATLAB functions/operations.

The performance of the rocket is described by the following equation:

$$\textbf{Height} = \textbf{60} + \textbf{2.13t}^2 - \textbf{0.0013t}^4 + \textbf{0.000034t}^{4.751}$$

The equation gives the height above the ground at time **t** in feet. The rocket's nose (reference point) is initially 60 ft above ground level.

The program should start at time t and end when the rocket hits the ground, or stop after 100 seconds, computing in two second increments. The program should then be modified to print the time when the rocket begins to fall and when it impacts.

Notes

Handle Graphics® and User Interfaces

5.1 Introduction and Objectives

The following chapter provides a brief introduction to Handle Graphics® and to developing your own GUIs in MATLAB®.

Upon completion of this chapter, the reader will be able to identify MATLAB graphics objects and their structure and operation and build a simple GUI using MATLAB's GUI Layout Editor (GUIDE).

5.2 Handle Graphics®

Handle Graphics are MATLAB's graphics objects and their properties. It is helpful for the user to be able to customize the previously provided MATLAB graphics. The key MATLAB Handle Graphics Objects and Operations are introduced in this section.

Each instance of an object is associated with a unique identifier called a handle. Using this handle, you can manipulate the characteristics or object properties of an existing graphics object. These are organized into a tree-structured hierarchy. This hierarchy is illustrated in Fig. 5.1.

Root is at the top of this hierarchy, the **figure** underneath, **axes** and similar qualities below that, and **image** qualities at the lowest level. To draw a **line** object, MATLAB needs an **axis** object to orient and provide a frame of reference to the line. The **axis**, in turn, needs a figure window to display the line.

The details on all the different types of graphics objects are found in the Graphics section of the MATLAB Help Documentation under **Graphics: Handle Graphics Objects: Types of Graphics.** The Handle Graphics Objects and Operations follow. The detailed manipulation of these properties is best covered later in this section.

Finding and Identifying Graphics Objects

allchild	Find all children of specified objects
ancestor	Find ancestor of graphics object
copyobj	Make copy of graphics object and its children
delete	Delete files or graphics objects
findall	Find all graphics objects (including hidden handles)
figflag	Test if figure is on screen

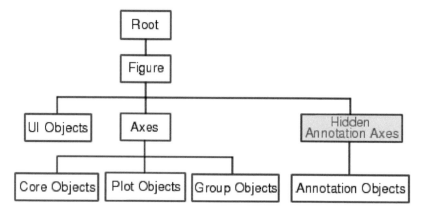

Fig. 5.1   Handle Graphics® Tree-structured Hierarchy.

findfigs	Display off-screen visible figure windows
findobj	Find objects with specified property values
gca	Get current axes handle
gcbo	Return object whose callback is currently executing
gcbf	Return handle of figure containing, callback object
gco	Return handle of current object
get	Get object properties
ishandle	True if value is valid object handle
set	Set object properties

Object Creation Functions

axes	Create axes object
figure	Create figure (graph) windows
hggroup	Create a group object
hgtransform	Create a group to transform
image	Create image (two-dimensional matrix)
light	Create light object (illuminates Patch and Surface)
line	Create line object (three-dimensional polylines)
patch	Create patch object (polygons)
rectangle	Create rectangle object (two-dimensional rectangle)
rootobject	List of root properties
surface	Create surface (quadrilaterals)
text	Create text object (character strings)
uicontextmenu	Create context menu (pop-up associated with object)

Plot Objects

areaseries	Property list
barseries	Property list
contourgroup	Property list
errorbarseries	Property list

lineseries	Property list
quivergroup	Property list
scattergroup	Property list
stairseries	Property list
stemseries	Property list
surfaceplot	Property list

Figure Windows

clc	Clear figure window
clf	Clear figure
close	Close specified window
closereq	Default close request function
drawnow	Complete any pending drawing
figflag	Test if figure is on screen
gcf	Get current figure handle
hgload	Load graphics object hierarchy from a FIG-file
hgsave	Save graphics object hierarchy to a FIG-file
newplot	Graphics M-file preamble for NextPlot property
opengl	Change automatic selection mode of OpenGL rendering
refresh	Refresh figure
saveas	Save figure or model to desired output format

Axes Operations

axis	Plot axis scaling and appearance
box	Display axes border
cla	Clear axes
gca	Get current axes handle
grid	Grid lines for two- and three-dimensional plots
ishold	Get the current hold state
makehgtform	Create a transform matrix

Operating on Object Properties

get	Get object properties
linkaxes	Synchronize limits of specified axes
linkprop	Maintain same value for corresponding properties
set	Set object properties

Let us next use Handle Graphics objects to modify the properties of an axis. First let us make an axis and then plot some data:

$\gg$ **x = 0:.01:10;**

$\gg$ **plot(x,sin(x).*x.$^\wedge$2);**

The result is shown in Fig. 5.2.

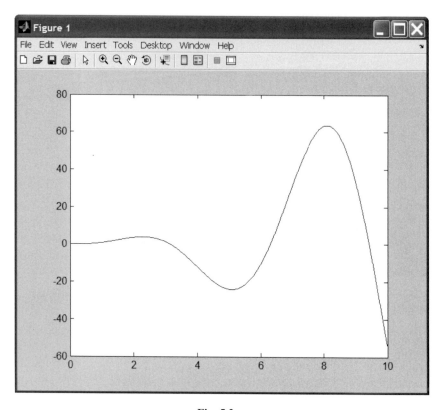

Fig. 5.2

The following command gets all the properties of the current axis:

```
≫ get(gca)
     ActivePositionProperty = outerposition
     ALim = [0 1]
     ALimMode = auto
     AmbientLightColor = [1 1 1]
     Box = on
     CameraPosition = [5 10 17.3205]
     CameraPositionMode = auto
     CameraTarget = [5 10 0]
     CameraTargetMode = auto
     CameraUpVector = [0 1 0]
     CameraUpVectorMode = auto
     CameraViewAngle = [6.60861]
     CameraViewAngleMode = auto
     CLim = [0 1]
     CLimMode = auto
```

Color = [1 1 1]
CurrentPoint=[(2 by 3) double array]
ColorOrder=[(7 by 3) double array]
DataAspectRatio = [5 70 1]
DataAspectRatioMode = auto
DrawMode = normal
FontAngle = normal
FontName = Helvetica
FontSize = [10]
FontUnits = points
FontWeight = normal
GridLineStyle = :
Layer = bottom
LineStyleOrder = -
LineWidth = [0.5]
MinorGridLineStyle = :
NextPlot = replace
OuterPosition=[(1 by 4) double array]
PlotBoxAspectRatio = [1 1 1]
PlotBoxAspectRatioMode = auto
Projection = orthographic
Position=[(1 by 4) double array]
TickLength = [0.01 0.025]
TickDir = in
TickDirMode = auto
Title = [153.004]
Units = normalized
View = [0 90]
XColor = [0 0 0]
XDir = normal
XGrid = off
XLabel = [154.004]
XAxisLocation = bottom
XLim = [0 10]
XLimMode = auto
XMinorGrid = off
XMinorTick = off
XScale = linear
XTick=[(1 by 6) double array]
XTickLabel =
     0
     2
     4
     6
     8
     10
XTickLabelMode = auto
XTickMode = auto

YColor = [0 0 0]
YDir = normal
YGrid = off
YLabel = [155.005]
YAxisLocation = left
YLim = [−60 80]
YLimMode = auto
YMinorGrid = off
YMinorTick = off
YScale = linear
YTick=[(1 by 8) double array]
YTickLabel =
        −60
        −40
        −20
          0
         20
         40
         60
         80
YTickLabelMode = auto
YTickMode = auto
ZColor = [0 0 0]
ZDir = normal
ZGrid = off
ZLabel = [156.005]
ZLim = [−1 1]
ZLimMode = auto
ZMinorGrid = off
ZMinorTick = off
ZScale = linear
ZTick = [−1 0 1]
ZTickLabel =
ZTickLabelMode = auto
ZTickMode = auto
BeingDeleted = off
ButtonDownFcn =
Children = [152.006]
Clipping = on
CreateFcn =
DeleteFcn =
BusyAction = queue
HandleVisibility = on
HitTest = on
Interruptible = on
Parent = [1]
Selected = off
SelectionHighlight = on

Tag =
Type = axes
UIContextMenu = []
UserData = []
Visible = on

Let us next flip the y axis so that negative values are at the top (see Fig. 5.3):

>> **set(gca,'ydir','reverse');**

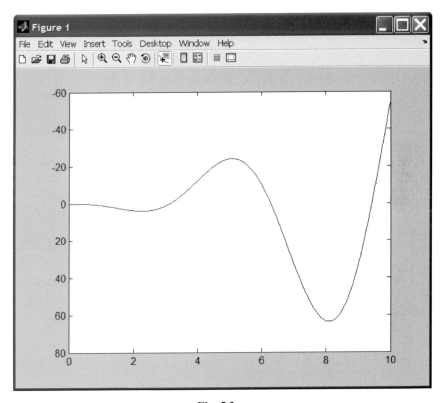

Fig. 5.3

Let us look up what options are provided for the **'tickDir'** property. To do this, we call **set** without giving the command a value. MATLAB will then print out all of the available options:

>> **set(gca,'tickDir')**
[{in} | out]

The tick marks are currently set to **'in'**. Now we will change the tick marks from **'in'** to **'out'** (see Fig. 5.4):

>> **set(gca,'tickDir','out')**

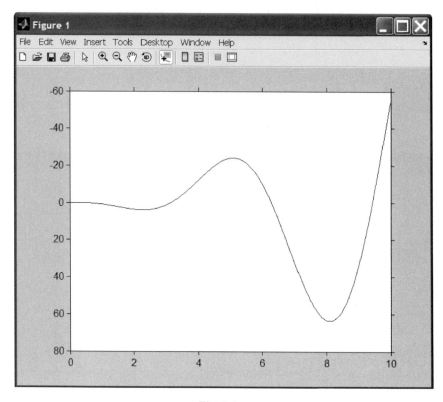

Fig. 5.4

Now we will put a text string on the graph:

>> **h = text(.5,.5,'my text');**

The results are shown in Fig. 5.5.

Now we need to move the position of the text string and change the font size and weight:

>> **set(h,'position',[5 −40],'fontsize',20,'fontweight','bold')**

The resulting figure is plotted next with the modified text string (see Fig. 5.6).

5.3   Graphical User Interface Development Environment

What is a graphical user interface (GUI)? A GUI is a user interface built with graphical objects—the components of the GUI—such as buttons, text fields, sliders, and menus. If the GUI is well designed, it should be intuitively obvious to the user how its components function. For example, when you

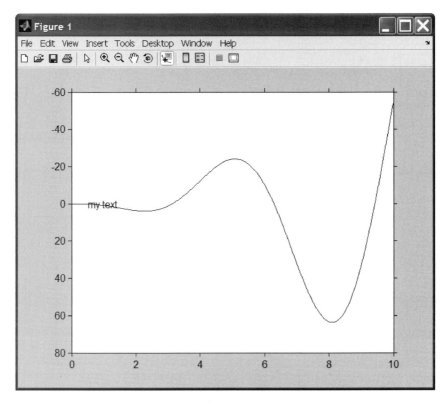

Fig. 5.5

move a slider, a value changes; when you click an OK button, your settings are applied and the dialog box is closed. Fortunately, most computer users are already familiar with GUIs and know how to use standard GUI components. By providing an interface between the user and the application's underlying code, GUIs enable the user to operate the application without knowing the commands that would be required by a command line interface. For this reason, applications that provide GUIs are easier to learn and use than those that are run from the command line. The sections that follow describe how to create GUIs with GUIDE. This includes laying out the components, programming them to do specific things in response to user actions, and saving and opening the GUI.

GUIDE, the MATLAB graphical user interface development environment, provides a set of tools for creating GUIs. GUIDE simplifies the process of creating GUIs by automatically generating the GUI M-file directly from your layout. GUIDE generates callbacks for each component in the GUI that requires a callback. Initially, GUIDE generates just a function definition line for each callback. You can add code to the callback to make it perform the operation you want. These tools greatly simplify the process of laying out and programming a GUI. This section introduces you to GUIDE and the layout tools it provides.

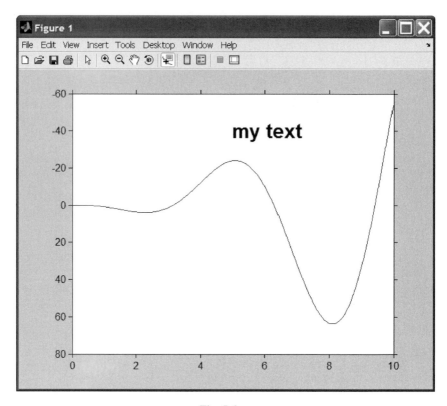

Fig. 5.6

The GUI Layout Editor is started from MATLAB using the command **guide**. Invoking **guide** displays the GUI Layout Editor opened to a new untitled FIG-file. By qualifying the command, **guide('filename.fig')** opens the FIG-file named **filename.fig**. You can even specify the path to a file not on your MATLAB path. The command **guide('figure_handles')** opens FIG-files in the Layout Editor for each existing figure listed in **figure_handles**. MATLAB copies the contents of each figure into the FIG-file, with the exception of axes children (i.e., image, light, line, patch, rectangle, surface, and text objects), which are not copied.

GUIDE provides several templates, which are simple examples that you can modify to create your own GUIs. The templates are fully functional GUIs: their callbacks are already programmed. You can view the code for these callbacks to see how they work and then modify the callbacks for your own purposes. You can access the templates in two ways.

1) Start GUIDE by entering **guide** at the MATLAB prompt.

2) If GUIDE is already open, select **New** from the **File** menu in the **Layout Editor.**

Starting GUIDE displays the **GUIDE Quick Start** dialog as shown in Fig. 5.7.

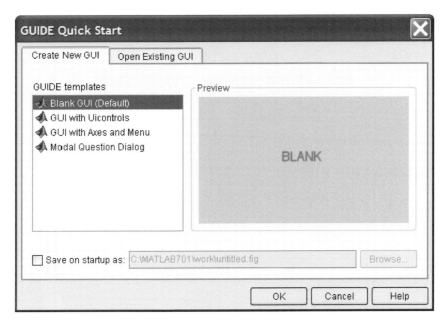

Fig. 5.7

5.4   Layout Editor

When you open a GUI in GUIDE, it is displayed in the Layout Editor (see Fig. 5.8), which is the control panel for all of the GUIDE tools. The Layout Editor enables you to lay out a GUI quickly and easily by dragging components, such as push buttons, pop-up menus, or axes, from the component palette into the layout area. Once you lay out your GUI and set each component's properties, using the tools in the Layout Editor, you can program the GUI with the M-file Editor. Finally, when you press the **Run** button on the toolbar, the functioning GUI appears outside the Layout Editor window.

This view has the option to show component names.

1) **Align Objects:** The **Alignment Tool** enables you to position objects with respect to each other and to adjust the spacing between selected objects. The specified alignment operations apply to all components that are selected when you press the **Apply** button. The **Alignment Tool** provides two types of alignment operations: 1) **align**, which aligns all selected components to a single reference line, and **distribute**, which spaces all selected components uniformly with respect to each other. Both types of alignment can be applied in the vertical and horizontal directions. Note that, in many cases, it is better to apply alignments independently to the vertical or to the horizontal using two separate steps.

2) **Menu Editor:** MATLAB enables you to create two kinds of menus: 1) **menu bar objects**, menus displayed on the figure **menubar**, and 2) **context menus**, menus that pop up when users right-click on graphics objects. You create both types of menus using the **Menu Editor**, which you can access from the **Menu Editor** item on the **Tool** menu and from the **Layout Editor** toolbar.

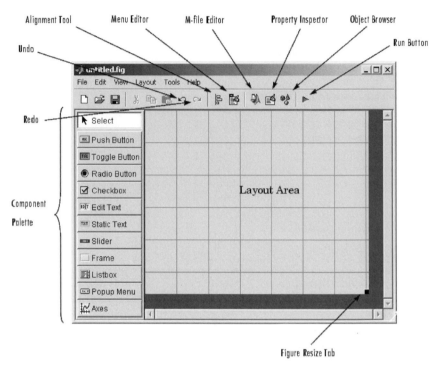

Fig. 5.8

3) **M-File Editor:** The **M-file Editor** creates a new M-file or opens an existing M-file in the MATLAB Editor/Debugger.

4) **Property Inspector:** The **Property Inspector** enables you to set the properties of the components in your layout. It provides a list of all settable properties and displays the current value. Each property in the list is associated with an editing device that is appropriate for the values accepted by the particular property; for example, a color picker to change the **BackgroundColor**, a pop-up menu to set **FontAngle**, and a text field to specify the **Callback** string.

5) **Object Browser:** The **Object Browser** displays a hierarchical list of the objects in the figure. In the following example, we will illustrate a figure object and its child objects.

6) **Run:** A GUI can be created to run simulations and plot the results.

The following **uicontrol objects** are available in the component palette.

1) **Push Buttons: Push Buttons** generate an action when clicked (e.g., an **OK** button may close a dialog box and apply settings). When you click the mouse on a **Push Button**, it appears depressed; when you release the mouse, the button appears raised, and its callback executes.

2) **Toggle Buttons: Toggle Buttons** generate an action and indicate a binary state (e.g., on or off). When you click on a **Toggle Button**, it appears depressed

and remains depressed until you release the mouse button, at which point the call-back executes. A subsequent mouse click returns the **Toggle Button** to the raised state and again executes its callback.

3) **Radio Buttons: Radio Buttons** are similar to **Check Boxes** (to follow) but are intended to be mutually exclusive within a group of related **Radio Buttons** (i.e., only one button is in a selected state at any given time). To activate a **Radio Button**, click the mouse on the object. The display indicates the state of the button.

4) **Check Boxes: Check Boxes** generate an action when checked and indicate their state as checked or not checked. **Check Boxes** are useful when providing the user with a number of independent choices that set a mode (e.g., display a toolbar or generate callback function prototypes).

5) **Edit Text: Edit Text** controls are fields that enable users to enter or modify text strings. Use **Edit Text** when you want text as input. The **String** property contains the text entered by the user.

6) **Static Text: Static Text** controls display lines of text. **Static Text** is typically used to label other controls, provide directions to the user, or indicate values associated with a slider. Users cannot change static text interactively, and there is no way to invoke the callback routine associated with it.

7) **Sliders: Sliders** accept numeric input within a specific range by enabling the user to move a sliding bar. Users move the bar by pressing the mouse button and dragging the slide, by clicking in the trough, or by clicking an arrow. The location of the bar indicates a numeric value.

8) **Frames: Frames** are boxes that enclose regions of a figure window. **Frames** can make a user interface easier to understand by visually grouping related controls. **Frames** have no callback routines associated with them and only uicontrols can appear within frames (axes cannot).

9) **List Boxes: List Boxes** display a list of items and enable users to select one or more items.

10) **Pop up Menus: Pop up Menus** open to display a list of choices when users click the arrow.

5.5   Property Inspector

The **Property Inspector** allows you to directly edit the properties of the selected object. In Fig. 5.9, you see the properties of a push button. The most used properties will be **String** (the label) and **Callback** (the action to perform). In Fig. 5.10, you see one way to edit the callback by right-clicking on the button and selecting **Callback.** This brings up the MATLAB Editor and places you at the entry point of the callback that was automatically created for you. Before you bring up the editor, make sure that you name your control properly, because the name will be used in the function that is created for you.

The GUI M-file generated by GUIDE controls the GUI and determines how it responds to a user's actions, such as pressing a push button or selecting a menu item. The M-file contains all the code needed to run the GUI, including the callbacks for the GUI components. While GUIDE generates the framework

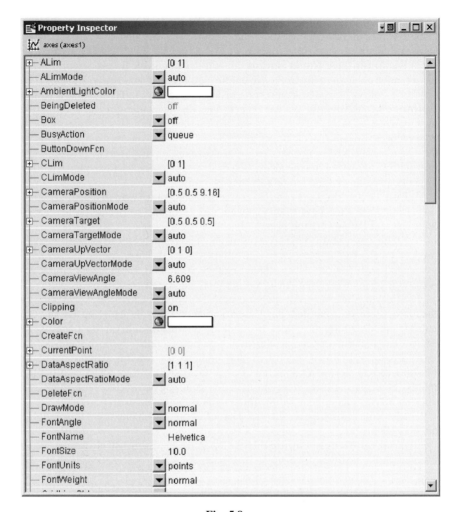

Fig. 5.9

for this M-file, you must program the callbacks to perform the functions you want them to.

The code shown as follows is a typical default callback that you can then fill in with your code.

% — Executes on button press in pushbutton1.
function pushbutton1_Callback(hObject, eventdata, handles)
% hObject handle to pushbutton1 (see GCBO)
% eventdata reserved — to be defined in a future version of MATLAB
% handles structure with handles and user data (see GUIDATA)

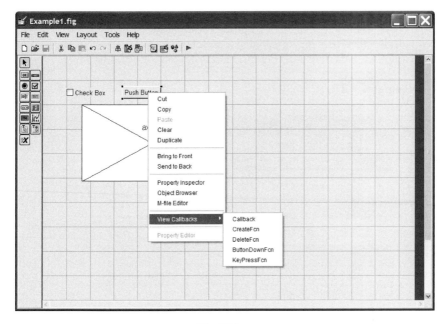

Fig. 5.10

If you would rather call some other function, edit the **Callback** property in the inspector to suit your needs. For example, making the **Callback** property **read**

myProgram('Push Button')

could be used to alert your **myProgram** that the button was pushed.

Although you can certainly use MATLAB's automatically created callbacks, it is advisable to put all of your analysis code in a separate file to maintain readability and reusability.

If we would like to create a GUI that has a graph and a push button that plots a sine curve when activated, we can proceed as follows.

1) Build the GUI with a graph and a push button. Also include a checkbox to set an option. Save it as (see Fig. 5.11)

Example1_GUI.

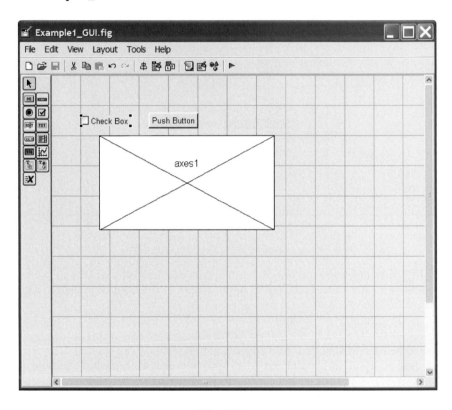

Fig. 5.11

2) Set the callback for the button to be (see Fig. 5.12)

example1('Button Pushed')

Fig. 5.12

3) Start a new M-file called **example1.m**. Load the GUI if no arguments are called or handle the argument if it is present (see Fig. 5.13).

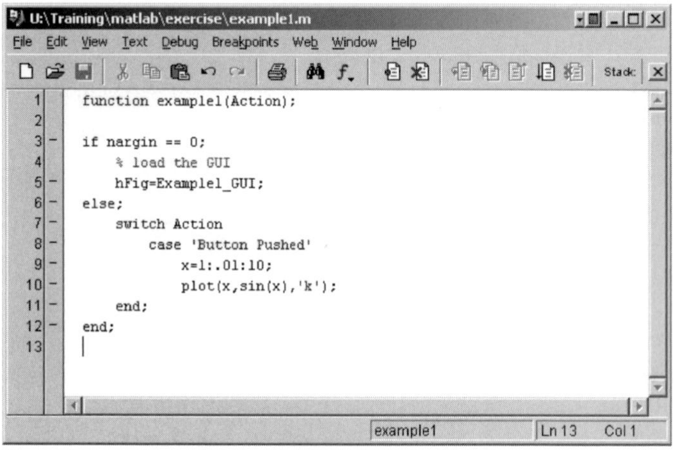

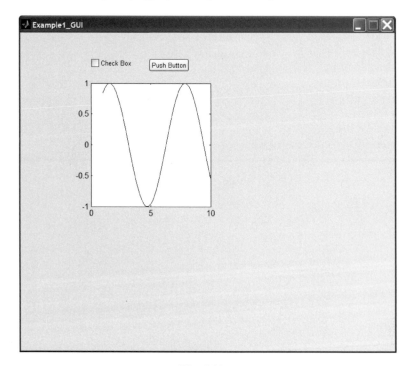

U:\Training\matlab\exercise\example1.m

File Edit View Text Debug Breakpoints Web Window Help

```
1    function example1(Action);
2
3    if nargin == 0;
4        % load the GUI
5        hFig=Example1_GUI;
6    else;
7        switch Action
8            case 'Button Pushed'
9                x=1:.01:10;
10               plot(x,sin(x),'k');
11           end;
12   end;
13
```

example1   Ln 13   Col 1

Fig. 5.13

4) Run the example by typing **example1** (see Fig. 5.14).

Fig. 5.14

5) Now check the status of the checkbox and do another plot if it is checked (see Fig. 5.15):

```
case 'Button Pushed'
  hCheckBox = findobj('tag','checkbox1');
  Selected = get(hCheckBox,'value');
  x = 1:.01:10;
  if Selected;
    plot(x,cos(x),'k',x,sin(x),'r');
  else;
    plot(x,sin(x),'k');
  end;
```

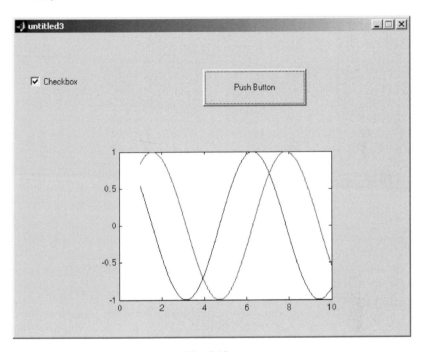

Fig. 5.15

5.6   Menu Editor

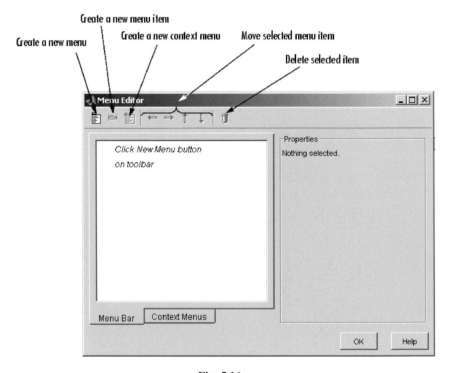

Fig. 5.16

Menus (as well as complete GUIs) can be created via MATLAB commands. However, it is easier to build them with the **Menu Editor** (see Fig. 5.16). Using the **New Menu** and **New Menu Item** buttons, you can create either standard windows-style menus or your own custom menus. Just like graphical objects, you interact with your code with callbacks. In Fig. 5.17, a callback is shown that tells the program to open a file.

Now implement the callback in the **example1.m** file:

case 'File Open'
  [file,path] = uigetfile;

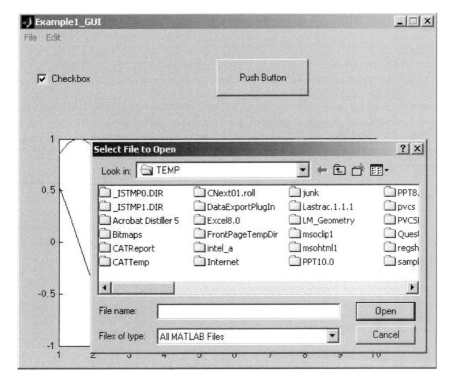

Fig. 5.17

In addition to standard menus, you can create context-sensitive menus that get activated when you right-click on the figure. One very useful feature is that you can change any menu at run time. This allows you to update the menu based on the program status. Items have a checked property that allows check marks to be placed next to them to indicate that an item is active.

Most of the options that you see in the property inspector can be changed at run time. Thus, a good knowledge of Handle Graphics will go a long way in helping you to customize your GUI and make a well-designed program.

5.7 Compiling a Stand-Alone Executable

MATLAB allows you to compile your programs into stand-alone C or C++ executables, complete with GUIs. There are too many options to cover here, but we can simply compile our **example1** program by issuing the following command at the MATLAB prompt:

≫ **mcc −B sgl example1**

This converts the M-file functions into C and compiles them, links them with the MATLAB libraries, and produces the **example1.exe** file. To run this from the MATLAB prompt, simply type **!example1**.

To install this program on PCs that do not have MATLAB, you need to do the following:

1) Copy the executable .fig files and any contents of the \bin directory to the new machine.

2) Copy any custom mex files that the program uses to the new machine.

3) Run the MATLAB Library installer on the new machine, which is located on the original computer at **$MATLAB\extern\lib\win32\mglinstaller.exe**. Here **$MATLAB** is the location of the MATLAB installation.

4) Add the . . .**\bin\win32** directory that was specified during the installation of the MATLAB Library to the system path.

5.8 Conclusion

This chapter was a brief introduction to the GUI creation capabilities of MATLAB. Only a small part of the capabilities were explored. MATLAB gives users the power to build high-quality programs driven by functional user interfaces very easily.

Practice Exercises

5.1 Assignment objective: Design a simple, graphically driven program that loads in data from a csv-formatted (data only) file and plots it on an axis. The filename should be read from an edit box and can be typed in or inserted from the result of **uigetfile** call. Your final program should resemble that shown in Fig. 5.18 for a data file of a sine wave.

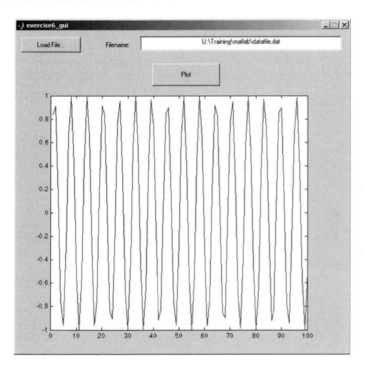

Fig. 5.18

Notes

Introduction to MATLAB® MEX-Files

6.1 Introduction and Objectives

This chapter covers some of the other data types that are available in MATLAB®. It also covers MEX-files. MEX-files enable routines written in other programming languages (such as FORTRAN) to interface with MATLAB.

Upon completion of this chapter, the reader will be able to identify other data types available in MATLAB and set dynamically linked subroutines using MEX-files.

6.2 Dynamically Linked Subroutines: MEX-Files

Although MATLAB is a complete, self-contained environment for programming and manipulating data, it is often useful to interact with data and programs external to the MATLAB environment. You can call your own C and FORTRAN subroutines from MATLAB as if they were built-in functions. MATLAB-callable C and FORTRAN programs are referred to as MEX-files. MEX-files are dynamically linked subroutines that the MATLAB interpreter can automatically load and execute.

MEX-files have several applications.

1) Large pre-existing FORTRAN and C programs can be called from MATLAB without being rewritten as MATLAB M–files.

2) Bottleneck computations (usually for-loops) that do not run fast enough in MATLAB can be recoded in C or FORTRAN and compiled for faster execution.

MEX-files are not appropriate for all applications. MATLAB is a high-productivity system that is designed to eliminate time-consuming, low-level programming in compiled languages like FORTRAN or C. In general, most programming should be done in MATLAB. Don't use the MEX facility unless your application requires it.

What kind of things can you do with MEX-files?

1) *You can build interfaces to existing libraries.* There are many tools available in the form of libraries. To make use of them, you don't have to abandon MATLAB and do all your work in C (or FORTRAN). Just write a MEX-file wrapper, link against the library, and you can make library calls within MATLAB.

2) *You can speed up your M-files.* Efficiently vectorized M-files run about as fast as if they were written in C (well, within a factor of 2 or 3). M-files that do a lot of looping can be significantly slower. Some problems can't readily be vectorized (usually problems where future results depend on past results), and these are ideal candidates for conversion into MEX-files.

6.2.1 Using MEX-Files

MEX-files are dynamically linked subroutines produced from C or FORTRAN source code; they behave just like M-files (files that end in .m) and built-in functions. To distinguish them from M-files on disk, MEX-files use the extension mex followed by the platform-specific identifier. Table 6.1 shows the platform-specific extensions for MEX-files.

To invoke a MEX-file on your disk, MATLAB looks through the list of directories on MATLAB's search path. It scans each directory in order looking for the first occurrence of the function with the filename extension mex or m. When it finds one, it loads the file and executes it. MEX-files behave just like M-files and built-in functions. Whereas M-files have a platform-independent extension, m, MATLAB identifies MEX-files by platform-specific extensions. MEX-files take precedence over M-files when like-named files exist in the same directory.

6.2.2 Matrix Data Structure

Before you can program MEX-files, you must become acquainted with the internal data structures or objects used by MATLAB. The MATLAB mathematics language works only with a single object type: the matrix. A matrix can be square or rectangular, complex or real, and full or sparse. Scalars, vectors, and text all are represented using the same structure.

Rather than directly manipulating the matrix data structure from C (which is possible) or from FORTRAN (which is not possible), the MEX-file interface library provides a set of access subroutines for manipulating matrices.

A matrix object contains the properties listed in Table 6.2.

If the storage class of a matrix is sparse, **pr** and **pi** have slightly different interpretations and the new properties—**nzmax, ir,** and **jc**—are relevant (see Table 6.3).

Table 6.1   MEX-file extensions for different computer systems

System type	MEX-file extension
Sun	Solaris.mexsol
HP-UX	.mexhpux
Linux	.mexglx
MacIntosh	.mexmac
Windows	.dll

Table 6.2 Matrix object properties

Property	Description
Name	Points to a character string array of length **mxMAXNAN** containing the null-terminated name of the matrix. **mxMAXNUM** is defined to be 20 in the file matrix.h. If the matrix is temporary (the result of an intermediate expression), the first character in the name is '\0'.
M	The number of rows in the matrix.
N	The number of columns in the matrix.
DisplayMode	Instructs MATLAB to either display the matrix in numeric form or to interpret the elements as ASCII values and display the matrix as a string, if the semicolon is omitted from the statement.
Storage	Indicates the storage class; that is, whether the matrix is full or sparse.
pr	Points to an array containing the real part of the matrix. The real part of the matrix consists of a length M*N (length is **nzmax** if the matrix is sparse), contiguous, singly subscripted, array of double-precision (64-bit) floating-point numbers. The elements of the matrix are stored using FORTRAN. The number of rows in the matrix is stored using FORTRAN's column-order convention (not C's row-wise convention).
pi	Points to an array containing the imaginary part of the matrix. If this pointer is NULL, there is no imaginary part, and the matrix is purely real.

6.2.3   C Language MEX-Files

MEX-files are built by combining your C source code with a set of routines provided in the MATLAB External Interface Library.

6.2.3.1   Directory organization.
A collection of files associated with the creation of C language MEX-files is located on the disk in the directory named **$MATLAB/extern**, where **$MATLAB** is the directory in which MATLAB is installed. Beneath this directory are three subdirectories into which the files are grouped:

$MATLAB/extern/include
$MATLAB/extern/lib
$MATLAB/extern/src

The /**include** subdirectory holds header files containing function declarations for all routines that you can access in the External Interface Library:

mex.h	**MEX-file function prototypes**
matrix.h	**Matrix access methods prototypes**

Table 6.3   Array lengths and pointers

Property	Description
Nzmax	The length of **ir**, **pr** and, if it exists, **pi**. It is the maximum possible number of nonzero elements in the sparse matrix.
pr	Points to the double-precision array of length **nzmax** containing the real parts of the nonzero elements in the matrix.
pi	NULL, if the matrix is real; otherwise it points to the double-precision array of length **nzmax** containing the imaginary parts of the nonzero elements in the matrix.
ir	Points to an integer array of length **nzmax** containing the row indices for the corresponding elements in **pr** and **pi**.
jc	Points to an integer array of length N+1 that contains column index information. For j in the range $0<=j<=N-1$, **jc[j]** is the index in **ir**, **pr**, (and **pi**, if it exists) of the first nonzero entry in the jth column, and **jc[j+1]-1** is the index of the last nonzero entry. As a result, **jc[N]** is also equal to **nnz**, the number of nonzero entries in the matrix. If **nnz** is less than **nzmax**, then more nonzero entries can be inserted in the matrix without allocating additional storage.

You must include **mex.h** in your MEX-file source files. Note that **mex.h** includes **matrix.h**, and so there is no need to include **matrix.h** explicitly.

The /**lib** subdirectory contains the object libraries used when linking your MEX-files.

The /**src** subdirectory contains source code examples of MEX-files.

6.2.3.2   C language MEX-file examples.

Example 1—a very simple example.   The following is an example of a C language MEX-file to write the string "Hello World" to the command window:

```
/*- - - - - - - - - - - - - -simpex.c- - - - - - - - - - - - - -*/
#include "mex.h"
void mexFunction( int nlhs, mxArray *plhs[], int nrhs,
const mxArray *prhs[] ) {
mexPrintf("Hello World\n");
}
/*- - - - - - - - - - - - - - - - - - - - - - - - - - - - - - - */
```

The important points to notice are as follows.

The entry point is called **mexFunction**. You can have any number of other functions, but you must have at least one named **mexFunction**.

The parameters **nlhs** and **nrhs** tell you how many left- and right-hand-side arguments you supply the MEX-function within MATLAB. They are just like **nargin** and **nargout**. The parameters **plhs** and **prsh** are arrays of matrix structures.

To compile and link this file, we issue the following command:

≫ **mex simpex.c**

6.2.3.3 Example 2—MEX-file speed advantages.
M-files that do a lot of looping can be significantly increased in speed by implementation as MEX-files. One example of this sort of problem is modeling a switch with hysteresis. For instance, you might want it to turn on if an input is above 0.7 but stay turned on until it drops below 0.5.

The M-file would look something like this:

function out=hyst(x,min_thresh, max_thresh)
x=x(:);
for n=2:length(x),
 if x(n) > max_thresh,
  x(n)=1;
 elseif x(n) < min_thresh,
  x(n)=0;
 else,
 x(n)=x(n − 1);
 end
end
out=x;

The MEX-file equivalent is only slightly longer.

```
/*- - - - - - - - - - - - - -hystmex.c- - - - - - - - - */

#include "mex.h"

void mexFunction( int nlhs,
                  mxArray *plhs[],
                  int nrhs,
                  const mxArray *prhs[] )

/* nlhs and nrhs tell you how many left and right hand   */
/* arguments you supply to the MEX-function              */
/* plhs and prsb are arrays of Matrix structures.        */

{
    double *pr_out, *pr_in, *on, *off;
    long i, len;
    mxArray *mpout;

    /* This is the gateway portion */
    if (nrhs < 3)
        mexErrMsgTxt("Not enough input arguments");

    len = mxGetN(prhs[0]) * mxGetM(prhs[0]); /*Get first argument size*/
    pr_in = mxGetPr(prhs[0]);  /* Get pointer to the matrix data */
    off = mxGetPr(prhs[1]);    /* These are the scalars, so the first */
    on = mxGetPr(prhs[2]);     /*  element of each is the on/off value */
mpout = mxCreateDoubleMatrix(len,1,mxREAL); /* Create a REAL matrix to return */
    pr_out = mxGetPr(mpout);               /* Get pointer to the matrix data */
    pr_out[0] = pr_in[0];
```

```
/* In this example this is the computation portion */
for (i = 1; i < len; i++)
{
    if (pr_in[i] < *off)
        pr_out[i] = 0;
    else if (pr_in[i] > *on)
        pr_out[i] = 1;
    else
        pr_out[i] = pr_out[i - 1];
}

plhs[0] = mpout;  /* Assign the first left-hand side */
                  /* argument to mpout                */

return;
}
/*- - - - - - - - - - - - - - - - - - - - - - - */
```

Compile **hystmex.c** by typing

≫ **mex hystmex.c**

How do they compare in speed? This test was run on a SGI Indigo 2 extreme:

x=rand(50000,1);
tic; y=hyst(x,.5,.7);toc
   elapsed_time = 4.490 % M-file output due to tic and toc calls

tic; y2=hystmex(x,.5,.7);toc
   elapsed_time = 0.0368% M-file output from tic and toc calls

Therefore, the MEX-file is over 122 times faster. Of course, if it takes you more time to write, debug, and compile the MEX-file than you save due to the increased speed, it isn't a net gain (this is often the case).

6.2.3.4 Example 3—three-body problem. Consider the M-file **yprime.m**, which contains the differential equations for a restricted three-body problem. The program **yprime.m** returns state derivatives, given state values and time, and can be integrated using the MATLAB function **ode23**. The MathWorks recommends that you use its M-file, FORTRAN, and C versions of this program to validate your compiler and MEX-file implementation process. The source for **yprime.m** is the following function:

```
function yp = yprime(t,y)

% Differential equation system for the restricted three body problem.
% Think of a small third body in orbit about the earth and moon.
% The coordinate system moves with the earth-moon system.
% The 1-axis goes through the earth and the moon.
% The 2-axis is perpendicular, in the plane of motion of the third body.
% The origin is at the center of gravity of the two heavy bodies.
% Let mu = the ratio of the mass of the moon to the mass of the earth.
% The earth is located at (-mu,0) and the moon at (1 - mu,0).
% y(1) and y(3) = coordinates of the third body.
```

```
% y(2) and y(4) = velocity of the third body.
%
% Copyright (c) 1984-98 by The MathWorks, Inc., modified R. Colgren
% All Rights Reserved.

mu = 1/82.45;
mus = 1 - mu;
r1 = norm([y(1)+ mu, y(3)]);   % Distance to the earth
r2 = norm([y(1) - mus, y(3)]); % Distance to the moon
yp(1) = y(2);
yp(2) = 2*y(4) + y(1) - mus*(y(1)+mu)/r1^3 - mu*(y(1) - mus)/r2^3;
yp(3) = y(4);
yp(4) = -2*y(2) + y(3) - mus*y(3)/r1^3 - mu*y(3)/r2^3;
```

The following statements show how this function is used:

yprimeout = yprime(1,[1 2 3 4])

The following is the C language MEX-file version of **yprime.m**:

```
/*=========================================================================
 *
 * YPRIME.C Sample .MEX file corresponding to YPRIME.M
 *          Solves simple 3 body orbit problem
 *          Modified function to demonstrate shared library calling
 *
 * The calling syntax for the mex function is:
 *
 *          [yp] = yprime(t, y)
 *
 *  You may also want to look at the corresponding M-code, yprime.m.
 *
 * This is a MEX-file for MATLAB.
 * Copyright 1984-2002 The MathWorks, Inc., modified R. Colgren
 *
 *=======================================================================*/
/* $Revision: 1.1.6.2.1 $ */

#include <math.h>
#include "mex.h"
#define EXPORT_FCNS
#include "shrhelp.h"

/* Input Arguments */

#define    T_IN   prhs[0]
#define    Y_IN   prhs[1]

/* Output Arguments */

#define    YP_OUT   plhs[0]

#if !defined(MAX)
#define    MAX(A, B)    ((A) > (B) ? (A) : (B))
#endif

#if !defined(MIN)
#define    MIN(A, B)    ((A) < (B) ? (A) : (B))
#endif
```

```
#define PI 3.14159265

static    double    mu = 1/82.45;
static    double    mus = 1 - 1/82.45;

EXPORTED_FUNCTION void yprimefcn(
                double yp[],
                double *t,
                double y[]
                )
{
    double r1,r2;

    r1 = sqrt((y[0] + mu)*(y[0]+mu)  + y[2]*y[2]);
    r2 = sqrt((y[0] -mus)*(y[0]-mus) + y[2]*y[2]);
    /* Print warning if dividing by zero. */
    if (r1 == 0.0 || r2 == 0.0 ){
      mexWarnMsgTxt("Division by zero!\n");
    }

    yp[0] = y[1];
    yp[1] = 2*y[3]+y[0]-mus*(y[0]+mu)/(r1*r1*r1)-mu*(y[0]-mus)/(r2*r2*r2);
    yp[2] = y[3];
    yp[3] = -2*y[1] + y[2] - mus*y[2]/(r1*r1*r1) - mu*y[2]/(r2*r2*r2);
    return;
}
EXPORTED_FUNCTION mxArray* better_yprime(
                double t,
                mxArray* y_in)
{
    double *yp;
    double *y;
    unsigned int m,n;
    mxArray* yp_out;

    m = mxGetM(y_in);
    n = mxGetN(y_in);
    if (!mxIsDouble(y_in) || mxIsComplex(y_in) ||
       (MAX(m,n) != 4) || (MIN(m,n) != 1)) {
        mexErrMsgTxt("YPRIME requires that Y be a 4 x 1 vector.");
    }
    /* Create a matrix for the return argument */
    yp_out = mxCreateDoubleMatrix(m, n, mxREAL);

    /* Assign pointers to the various parameters */
    yp = mxGetPr(yp_out);

    y = mxGetPr(y_in);

    /* Do the actual computations in a subroutine */
    yprimefcn(yp,&t,y);
    return yp_out;
}

void mexFunction( int nlhs, mxArray *plhs[],
                  int nrhs, const mxArray*prhs[] )
```

```
{
    double *yp;
    double *t,*y;
    unsigned int m,n;

    /* Check for proper number of arguments */

    if (nrhs != 2) {
      mexErrMsgTxt("Two input arguments required.");
    } else if (nlhs > 1) {
      mexErrMsgTxt("Too many output arguments.");
    }
    /* Check the dimensions of Y. Y can be 4 X 1 or 1 X 4. */

    m = mxGetM(Y_IN);
    n = mxGetN(Y_IN);
    if (!mxIsDouble(Y_IN) || mxIsComplex(Y_IN) ||
        (MAX(m,n) ! = 4) || (MIN(m,n) ! = 1)) {
      mexErrMsgTxt("YPRIME requires that Y be a 4 x 1 vector.");
    }

    /* Create a matrix for the return argument */
    YP_OUT = mxCreateDoubleMatrix(m, n, mxREAL);

    /* Assign pointers to the various parameters */
    yp = mxGetPr(YP_OUT);

    t = mxGetPr(T_IN);
    y = mxGetPr(Y_IN);

    /* Do the actual computations in a subroutine */
    yprimefcn(yp,t,y);
    return;

}
```

To compile and link this file, we issue the following command:

> **mex yprime.c**

6.2.3.5 MEX-file details.   The source file for a MEX-file consists of two distinct parts: 1) a computational routine that contains the code for performing the actual numeric computation and 2) a gateway routine that interfaces the computational routine with MATLAB.

The parameters **nlhs** and **nrhs** contain the number of left-hand-side and right-hand-side arguments, respectively, with which the MEX-function is called. The parameter **prhs** is a length **nrhs** array of pointers to the right-hand-side matrices; **plhs** is a length **nlhs** array where your C function must put pointers for the returned left-hand-side matrices.

For example, from our previous yprime example we could write

> **x = yprime(1, [1 2 3 4]);**

At the MATLAB prompt, the MATLAB interpreter calls **mexFunction** with the arguments

nlhs = 1
nrhs = 2

plhs = "pointer to" NULL
prhs = "first item in array pointer points to" [1]
       "second item in array pointer points to"
       [1 2 3 4]

Because there are two right-hand arguments to **yprime**, **nrhs** is 2. The parameter **prhs[0]** points to a matrix object containing the **scalar** [1], whereas **prhs[1]** points to the matrix containing [1 2 3 4].

Note that in the C language matrix dimensions start at 0, not 1 as in FORTRAN. The caller is expecting **yprime** to return a single matrix. It is the responsibility of the gateway routine to create an output matrix (using **mxCreate DoubleMatrix**) and to store a pointer to that matrix into **plhs[0]**.

The gateway routine should dereference and validate the input arguments and call **mexErrMsgTxt** if anything is amiss.

The gateway routine must call **mxCreateDoubleMatrix**, **mxCreateSparse**, or **mxCreateString** to create matrices of the required sizes in which to return the results. The return values from these calls should be assigned to the appropriate elements of **plhs**.

The gateway routine may call **mxCalloc** to allocate temporary work arrays if needed.

Finally, the gateway routine should call the computational routine.

6.2.3.6 Debugging C MEX-file.   You cannot use a debugger on MEX-files invoked directly from MATLAB. There are two options available: 1) use the old standby printf statement and 2) generate a stand-alone executable image that can be tested outside MATLAB.

6.2.3.7 FORTRAN language MEX-file example.   Consider the same M-file **yprime.m** as in the C language example, which contains the differential equations for a restricted three-body problem. The program **yprime.m** returns state derivatives, given state values and time, and can be integrated using the MATLAB function **ode23**. The source for **yprime.m**   was   the   previous MATLAB function **yp = yprime(t,y)**.

The following is the FORTRAN language gateway MEX-file version of **yprime.m** called **yprimeg.f**.

```
C=================================================================
C YPRIMEG.FOR - Gateway function for YPRIME.FOR
C
C     This is an example of the FORTRAN code required for
C     interfacing a .MEX file to MATLAB.
C
C     This subroutine is the main gateway to MATLAB. When a
C     MEX function is executed MATLAB calls the MEXFUNCTION
C     subroutine in the corresponding MEX file.
C
C     Copyright 1984-2004 The MathWorks, Inc., modified R. Colgren
C     $Revision: 1.9.2.1.1 $
C=================================================================
```

```
C
      SUBROUTINE MEXFUNCTION(NLHS, PLHS, NRHS, PRHS)
C
C - - - - - - - - - - - - - - - - - - - - - - - - - - - - - - - - - -
C     (pointer) Replace integer by integer*8 on 64-bit platforms
C
      INTEGER PLHS(*), PRHS(*)
C
C - - - - - - - - - - - - - - - - - - - - - - - - - - - - - - - - - -
C
      INTEGER NLHS, NRHS
C
C - - - - - - - - - - - - - - - - - - - - - - - - - - - - - - - - - -
C     (pointer) Replace integer by integer*8 on 64-bit platforms
C
      INTEGER MXCREATEDOUBLEMATRIX, MXGETPR
C
C - - - - - - - - - - - - - - - - - - - - - - - - - - - - - - - - - -
C
      INTEGER MXGETM, MXGETN
C
C KEEP THE ABOVE SUBROUTINE, ARGUMENT, AND FUNCTION DECLARATIONS FOR USE
C IN ALL YOUR FORTRAN MEX FILES.
C - - - - - - - - - - - - - - - - - - - - - - - - - - - - - - - - - -
C
C - - - - - - - - - - - - - - - - - - - - - - - - - - - - - - - - - -
C     (pointer) Replace integer by integer*8 on 64-bit platforms
C
      INTEGER YPP, TP, YP
C
C - - - - - - - - - - - - - - - - - - - - - - - - - - - - - - - - - -
C
      INTEGER M, N
      REAL*8 RYPP(4), RTP, RYP(4)
C
C CHECK FOR PROPER NUMBER OF ARGUMENTS
C
      IF (NRHS .NE. 2) THEN
        CALL MEXERRMSGTXT('YPRIME requires two input arguments')
      ELSEIF (NLHS .GT. 1) THEN
        CALL MEXERRMSGTXT('YPRIME requires one output argument')
      ENDIF
C
C CHECK THE DIMENSIONS OF Y.  IT CAN BE 4 X 1 OR 1 X 4.
C
      M = MXGETM(PRHS(2))
      N = MXGETN(PRHS(2))
C
      IF ((MAX(M,N) .NE. 4) .OR. (MIN(M,N) .NE. 1)) THEN
        CALL MEXERRMSGTXT('YPRIME requires that Y be a 4 x 1 vector')
      ENDIF
C
C CREATE A MATRIX FOR RETURN ARGUMENT
C
      PLHS(1) = MXCREATEDOUBLEMATRIX(M,N,0)
C
C ASSIGN POINTERS TO THE VARIOUS PARAMETERS
C
```

```
      YPP = MXGETPR(PLHS(1))
C
      TP = MXGETPR(PRHS(1))
      YP = MXGETPR(PRHS(2))
C
C COPY RIGHT HAND ARGUMENTS TO LOCAL ARRAYS OR VARIABLES
      CALL MXCOPYPTRTOREAL8(TP, RTP, 1)
      CALL MXCOPYPTRTOREAL8(YP, RYP, 4)
C
C DO THE ACTUAL COMPUTATIONS IN A SUBROUTINE USING
C    CREATED ARRAYS.
C
      CALL YPRIME(RYPP,RTP,RYP)
C
C COPY OUTPUT WHICH IS STORED IN LOCAL ARRAY TO MATRIX OUTPUT
      CALL MXCOPYREAL8TOPTR(RYPP, YPP, 4)
C
      RETURN
      END
```

The following is the computational portion of the FORTRAN MEX-file called **yprime.f**.

```
C=========================================================
C The actual YPRIME subroutine in FORTRAN
C
C Copyright 1984-2000 The MathWorks, Inc., modified R. Colgren
C $Revision: 1.4.1 $
C=========================================================
C
      SUBROUTINE YPRIME(YP, T, Y)
      REAL*8 YP(4), T, Y(4)
C
      REAL*8 MU, MUS, R1, R2
C
      MU = 1.0/82.45
      MUS = 1.0 - MU
C
      R1 = SQRT((Y(1) +MU)**2 + Y(3)**2)
      R2 = SQRT((Y(1) -MUS)**2 + Y(3)**2)
C
      YP(1) = Y(2)
      YP(2) = 2*Y(4) + Y(1) - MUS*(Y(1)+MU)/(R1**3) -
     & MU*(Y(1) -MUS)/(R2**3)
C
      YP(3) = Y(4)
      YP(4) = -2*Y(2) + Y(3) - MUS*Y(3)/(R1**3) -
     & MU*Y(3)/(R2**3)
C
      RETURN
      END
```

To compile and link these files, we use the following command:

>> **mex yprime.f yprimeg.f**

This command carries out the necessary steps to create the MEX-file.

6.2.3.8 FORTRAN MEX-file details.   The source file for a MEX-file consists of two distinct parts: 1) a computational routine that contains the code for performing the actual numeric computation and 2) a gateway routine that interfaces the computational routine with MATLAB.

The computational and gateway routines must be split into separate files to "trick" FORTRAN compilers into allowing addresses to be treated as integer variables (pointers).

A pointer is a C language data type. Because MATLAB is written in C, pointers are used extensively inside MATLAB.

In FORTRAN, when a program calls a subroutine, arrays are passed by reference (by address). Rather than passing a complete copy of an array to a subroutine, only a copy of the address of the first element is passed. C pointers are like FORTRAN array names in that they point (i.e., refer) to objects rather than being the objects themselves. The entry point to the gateway subroutine must be named **mexFunction** and have the following parameters:

SUBROUTINE MEXFUNCTION(NLHS, PLHS, NRHS, PRHS)
C---

INTEGER*4 PLHS(*), PRHS(*)
C---

INTEGER NLHS, NRHS

The parameters **nlhs** and **nrhs** contain the number of left-hand-side and right-hand-side arguments, respectively, with which the MEX-function is called. The parameter **prhs** is a length **nrhs** array of pointers to the right-hand-side matrices. The parameter **plhs** is a length **nlhs** array where your C function must put pointers for the returned left-hand-side matrices.

For example, from our previous **yprime** example we could write

≫ x=yprime(1, [1 2 3 4]);

At the MATLAB prompt, the MATLAB interpreter calls **mexFunction** with arguments:

```
nlhs = 1
nrhs = 2
plhs = "pointer to" NULL
prhs = "first item in array pointer points to" [1]
       "second item in array pointer points to" [1 2 3 4]
```

Because there are two right-hand arguments to **yprime**, **nrhs** is **2**. The parameter **prhs[0]** points to a matrix object containing the scalar **[1]**, whereas **prhs[1]** points to the matrix containing **[1 2 3 4]**. Note that in the C language matrix dimensions start at 0, not 1 as in FORTRAN. The caller is expecting **yprime** to return a single matrix.

It is the responsibility of the gateway routine to create an output matrix (using **mxCreateDoubleMatrix**) and to store a pointer to that matrix into **plhs[0]**.

The gateway routine should dereference and validate the input arguments and call **mexErrMsgTxt** if anything is amiss.

The gateway routine must call **mxCreateDoubleMatrix**, **mxCreateSparse**, or **mxCreateString** to create matrices of the required sizes in which to return the results. The return values from these calls should be assigned to the appropriate elements of **plhs**.

The gateway routine may call **mxCalloc** to allocate temporary work arrays if needed.

Finally, the gateway routine should call the computational routine.

6.2.3.9 Debugging a FORTRAN MEX-file.   You cannot use a debugger on MEX-files invoked directly from MATLAB. There are two options available: 1) use the old standby write statement and 2) generate a stand-alone executable image that can be tested outside MATLAB.

6.3   MATLAB® Engine Library

The MATLAB engine library is a set of routines that allows you to call MATLAB from your C or FORTRAN programs (i.e., MATLAB as a subroutine) to use its mathematics and graphics routines.

Using this method, MATLAB is run in the background as a separate process and offers several advantages.

1) It can run on your machine or any other computer on the network, including machines of different architectures.

2) Instead of requiring MATLAB to be linked to your code ($>$3 meg), only a small communications library is needed.

Unfortunately, there is also a disadvantage. This method runs much slower than if it was all coded in FORTRAN or C.

6.3.1   Engine Library Routines

The engine library contains nine routines. Their names all begin with the prefix **eng**. Tables 6.4 and 6.5 list these routines.

Table 6.4   C engine routines

Function	Purpose
engOpen	Start up MATLAB engine
engClose	Shut down MATLAB engine
engGetVariable	Get a MATLAB array from the MATLAB engine
engPutVariable	Send a MATLAB array to the MATLAB engine
engEvalString	Execute a MATLAB command
engOutputBuffer	Create a buffer to store MATLAB text output
engOpenSingleUse	Start a MATLAB engine session for single, nonshared use
engGetVisible	Determine visibility of MATLAB engine session
engSetVisible	Show or hide MATLAB engine session

Table 6.5 FORTRAN engine routines

Function	Purpose
engOpen	Start up MATLAB engine
engClose	Shut down MATLAB engine
engGetVariable	Get a MATLAB array from the MATLAB engine
engPutVariable	Send a MATLAB array to the MATLAB engine
engEvalString	Execute a MATLAB command
engOutputBuffer	Create a buffer to store MATLAB text output

On UNIX, the engine library communicates with the MATLAB engine using pipes and, if needed, **rsh** for remote execution. On Microsoft Windows, the engine library communicates with MATLAB using a Component Object Model interface. **COM** and **DDE Support** contain a detailed description of **COM**.

6.3.2 Engine Library C Example

This program, **engtest1.c**, illustrates how to call the engine functions from a stand-alone C program. Additional engine examples are located in the eng_mat directory.

```
/*=====================================================================
 *     engtest1.c
 *
 *     This is a simple program that illustrates how to call the
 *     MATLAB Engine functions from a C program for Windows
 *
 *     The example starts a MATLAB engine process, then calculates the
 *     Position vs. Time for a falling object. Next, this example
 *     sends a 3-by-2 real matrix to it, computes the eigenvalues of
 *     the matrix multiplied by its transpose, gets the matrix back
 *     to the C program, and prints out the second eigenvalue.
 *
 *     Copyright 1984-2003 The MathWorks, Inc., modified R. Colgren
 *=====================================================================
 */

/* $Revision: 1.10.4.1 $ */

#include <windows.h>
#include <stdlib.h>
#include <stdio.h>
#include <string.h>
#include "engine.h"

#define BUFSIZE 256

static double Areal[6] = { 1, 2, 3, 4, 5, 6 };

int PASCAL WinMain (HINSTANCE   hInstance,
                    HINSTANCE   hPrevInstance,
                    LPSTR       lpszCmdLine,
                    int         nCmdShow)
{
    Engine *ep;
    mxArray *T = NULL, *a = NULL, *d = NULL;
    char buffer[BUFSIZE+1];
```

```
double *Dreal, *Dimag;
double time[10] = { 0, 1, 2, 3, 4, 5, 6, 7, 8, 9 };

/*
 * Start the MATLAB engine
 */
if (!(ep = engOpen(NULL))) {
    MessageBox ((HWND)NULL, (LPSTR)"Can't start MATLAB engine",
        (LPSTR) "Engwindemo.c", MB_OK);
    exit(-1);
}

/*
 * PART I
 *
 * For the first half of this demonstration, we will send data
 * to MATLAB, analyze the data, and plot the result.
 */

/*
 * Create a variable from our data
 */
T = mxCreateDoubleMatrix(1, 10, mxREAL);
memcpy((char *) mxGetPr(T), (char *) time, 10*sizeof(double));

/*
 * Place the variable T into the MATLAB workspace
 */
engPutVariable(ep, "T", T);

/*
 * Evaluate a function of time, distance = (1/2)g.*t.^2
 * (g is the acceleration due to gravity)
 */
engEvalString(ep, "D = .5.*(-9.8).*T.^2;");

/*
 * Plot the result
 */
engEvalString(ep, "plot(T,D);");
engEvalString(ep, "title('Position vs. Time for a falling object');");
engEvalString(ep, "xlabel('Time (seconds)');");
engEvalString(ep, "ylabel('Position (meters)');");

/*
 * PART II
 *
 * For the second half of this demonstration, we will create another mxArray
 * put it into MATLAB and calculate its eigenvalues
 */

a = mxCreateDoubleMatrix(3, 2, mxREAL);
memcpy((char *) mxGetPr(a), (char *) Areal, 6*sizeof(double));
engPutVariable(ep, "A", a);

/*
 * Calculate the eigenvalue
 */
engEvalString(ep, "d = eig(A*A')");

/*
 * Use engOutputBuffer to capture MATLAB output. Ensure first that
 * the buffer is always NULL terminated.
 */
buffer[BUFSIZE] = '\0';
engOutputBuffer(ep, buffer, BUFSIZE);

/*
 * the evaluate string returns the result into the
 * output buffer.
 */
```

```
    engEvalString(ep, "whos");
    MessageBox ((HWND)NULL, (LPSTR)buffer, (LPSTR) "MATLAB - whos", MB_OK);

    /*
     * Get the eigenvalue mxArray
     */
    d = engGetVariable(ep, "d");
    engClose(ep);

    if (d == NULL) {
            MessageBox ((HWND)NULL, (LPSTR)"Get Array Failed",
(LPSTR)"Engwindemo.c", MB_OK);
        }
    else {
        Dreal = mxGetPr(d);
        Dimag = mxGetPi(d);
        if (Dimag)
            sprintf(buffer,"Eigenval 2: %g+%gi",Dreal[1],Dimag[1]);
        else
            sprintf(buffer,"Eigenval 2: %g",Dreal[1]);
        MessageBox ((HWND)NULL, (LPSTR)buffer, (LPSTR)"Engwindemo.c", MB_OK);
        mxDestroyArray(d);
    }
    /*
     * We're done! Free memory, close the MATLAB engine and exit.
     */
    mxDestroyArray(T);
    mxDestroyArray(a);

    return(0);
}
```

To compile this code we use

**cc engtest1.c -o engtest1 - I$MATLAB/extern/include
$MATLAB/extern/lib/$ARCH/libmat.a -lm**

This creates the executable file **engtest1**.

6.3.3 Engine Library FORTRAN Example

The following is an example written in FORTRAN that illustrates how to call the engine functions from a stand-alone FORTRAN program.

```
C=======================================================
C     engtest2.f
C
C     This program illustrates how to call the MATLAB
C     Engine functions from a Fortran program.
C
C     Copyright (c) 1984-2000 by The MathWorks, Inc.
C     Modified R. Colgren
C     $Revision: 1.1.4.7.1 $
C=======================================================

      program main
      integer engOpen, engClose, engEvalString
      integer engGetVariable, engPutVariable
      integer mxGetPr, mxCreateDoubleMatrix
```

```fortran
      integer ep, T, D
      double precision time(10), dist(10)
      integer temp, status
      data time / 1.0, 2.0, 3.0, 4.0, 5.0, 6.0, 7.0, 8.0,
                  9.0, 10.0 /

      ep = engOpen('matlab')

      if (ep .eq. 0) then
         write(6,*) 'Can''t start MATLAB engine'
         stop
      endif

      T = mxCreateDoubleMatrix(1, 10, 0)
      call mxCopyReal8ToPtr(time, mxGetPr(T), 10)
C
C     Place the variable T into the MATLAB workspace.
C
      status = engPutVariable(ep, 'T', T)

      if (status .ne. 0) then
         write(6,*) 'engPutVariable failed'
         stop
      endif
C
C     Evaluate a function of time, distance = (1/2)g.*t.^2
C     (g is the acceleration due to gravity).
C
      if (engEvalString(ep, 'D = .5.*(-9.8).*T.^2;') .ne. 0)
     then
         write(6,*) 'engEvalString failed'
         stop
      endif


C
C     Plot the result.
C
      if (engEvalString(ep, 'plot(T,D);') .ne. 0) then
         write(6,*) 'engEvalString failed'
         stop
      endif

      if (engEvalString(ep, 'title(''Position vs. Time'')')
     .ne. 0) then
         write(6,*) 'engEvalString failed'
         stop
      endif

      if (engEvalString(ep, 'xlabel(''Time (seconds)'')')
     .ne. 0) then
         write(6,*) 'engEvalString failed'
         stop
      endif
C
```

```fortran
      if (engEvalString(ep, 'ylabel(''Position (meters)'')')
     .ne. 0) then
         write(6,*) 'engEvalString failed'
         stop
      endif
C
C     Read from console to make sure that we pause long
C     enough to be able to see the plot.
C
      print *, 'Type 0 <return> to Exit'
      print *, 'Type 1 <return> to continue'

      read(*,*) temp

      if (temp.eq.0) then
         print *, 'EXIT!'
         stop
      end if

      if (engEvalString(ep, 'close;') .ne. 0) then
         write(6,*) 'engEvalString failed'
         stop
      endif

      D = engGetVariable(ep, 'D')
      call mxCopyPtrToReal8(mxGetPr(D), dist, 10)
      print *, 'MATLAB computed the following distances:'
      print *, ' time(s) distance(m)'
      do 10 i=1,10
         print 20, time(i), dist(i)
20       format(' ', G10.3, G10.3)
10    continue

      call mxDestroyArray(T)
      call mxDestroyArray(D)
      status = engClose(ep)

      if (status .ne. 0) then
         write(6,*) 'engClose failed'
         stop
      endif

      stop
      end
```

To compile this, we would type the following command:

**F77 engtest2.f -o engtest2 -I$MATLAB/extern/include
$MATLABIextern/lib/$ARCH/1ibmat.a -lm**

This creates the executable file named **engtest2**.
This concludes our discussion of MATLAB MEX-files.

6.4 Conclusion

This chapter introduced the reader to some of the other data types available in MATLAB. It also introduced the reader to MEX-files and methods for calling MATLAB from C and FORTRAN programs.

Practice Exercises

6.1 Write a simple MEX-file in C that gets from the user a number, generates the square of that number, then displays the resulting square of that number back to the user. To set up a C compiler to use, if not already accomplished, type in the command

$\gg$ **mex -setup**

You will see something similar to the following request:

Please choose your compiler for building external interface (MEX) files:

Would you like mex to locate installed compilers [y]/n? **y**

(Note that you should select y to have MATLAB find all compilers.)
Select a compiler:

[1] Digital Visual FORTRAN version 6.6 in C:\msdev
[2] Lcc C version 2.4.1 in C:\PROGRAM FILES\MATLAB\R2006b\sys\lcc
[3] Microsoft Visual C/C++ version 7.1 in C:\Program Files\Microsoft Visual Studio
[0] None

(Note that here we select **2** to use the MATLAB provided LCC C compiler.)

Compiler: **2**

Then verify your choices to proceed with the assignment.

Notes

Basic Simulink®

Brief Introduction to Simulink® and Stateflow®

7.1 Introduction and Objectives

This chapter introduces some of the graphics modeling capabilities of Simulink® and Stateflow® software. It shows the reader how to open, execute, and modify parameters within these models from the MATLAB® **Command Window** and from MATLAB M-files.

Upon completion of this chapter, the reader will be able to identify graphics modeling capabilities of Simulink and Stateflow software, open and close Simulink and Stateflow models from the MATLAB **Command Window**, execute and simulate systems already implemented in Simulink and Stateflow software from the **Command Window**, and modify Simulink and Stateflow model parameters from MATLAB.

7.2 Simulink®

Simulink uses block diagrams to represent dynamic systems. Defining a system is much like drawing a block diagram. Blocks are copied from a library of blocks.

Typing the command **simulink** opens the standard block library. The standard block library is divided into several subsystems, grouping blocks according to their behavior (see Fig. 7.1).

Examples using the standard block libraries provided with Simulink are shown next. They include sources (which generate input signals), sinks (places where output data can be stored), linear blocks, nonlinear blocks, discrete blocks, and connections. The user can also construct block libraries.

7.3 Van der Pol Equation

This system models a second-order nonlinear system. A description of the system can be found on the diagram. This example demonstrates the ability of MATLAB to control the simulation of a system.

We start by loading the system:

>> **vdp;**

MATLAB will next simulate the system for 30 seconds and plot the time-varying behavior of X1 and X2 (see Fig. 7.2).

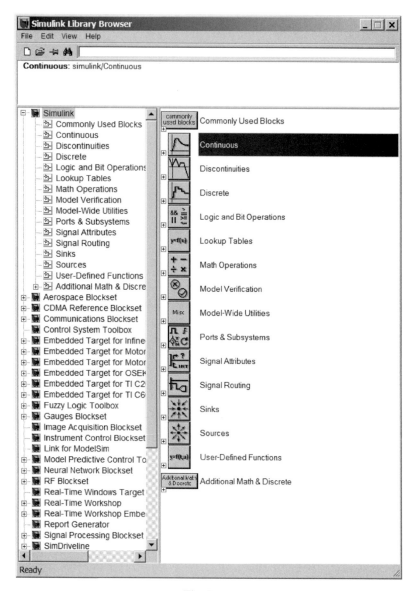

Fig. 7.1

> $\gg$ [t,x] = sim('vdp',30);

To generate the time history plots, type the following MATLAB commands (see Fig. 7.3):

> $\gg$ clf;
> $\gg$ subplot(211);
> $\gg$ title('The State Variables of the Van der Pol System');

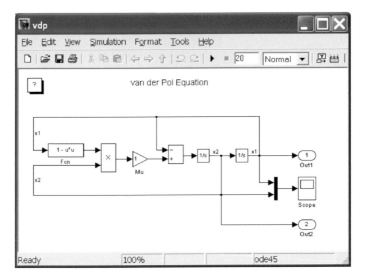

Fig. 7.2

```
>> plot(t,x(:,2));
>> ylabel('X1');
>> subplot(212);
>> plot(t,x(:,1));
>> ylabel('X2');
>> xlabel('Time in seconds');
>> subplot;
```

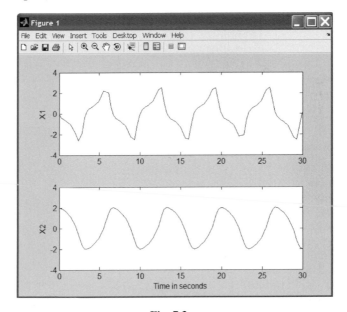

Fig. 7.3

Plotting X1 against X2 results in the phase-plane diagram (see Fig. 7.4):

```
>> clf;
>> plot(x(:,1),x(:,2));
>> title('The Phase Behavior of the Van der Pol System');
>> xlabel('X1'); ylabel('X2');
```

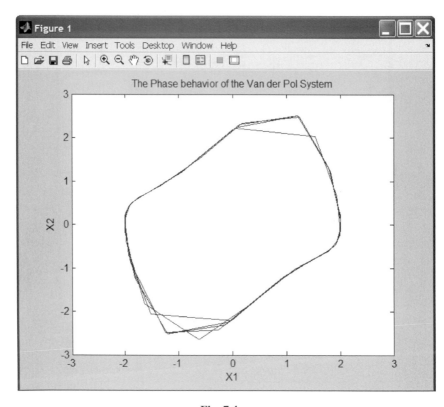

Fig. 7.4

MATLAB will now run 10 simulations, varying the fraction of X2 that is subtracted from X1, using the following M-file:

```
% M-file for the Van der Pol Simulation
time = 20;
iterations = 10;
step_size = 0.2;
x = ones(time/step_size + 1,iterations);
y = ones(time/step_size + 1,iterations);
set_param('vdp/Mu','Gain','a');
for n = 1:iterations
    a = 1 - n/iterations;
    [t,sf] = euler('vdp',time,[0.25,0.0],[5,step_size,step_size]);
```

```
    x(:,n) = sf(:,2);
    y(:,n) = sf(:,1);
end
clf;
mesh(x);
mesh(y);
title('The Effect of Reducing the Negative Feedback of X2');
```

Note that in the next version of MATLAB the command **euler** will become obsolete and may be eliminated. The MathWorks recommends using **sim** as a substitute for **euler** (see Fig. 7.5).

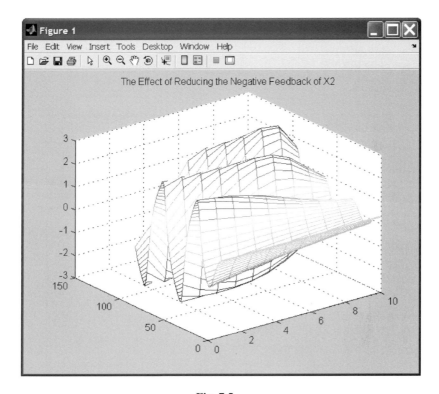

Fig. 7.5

Slices of these surfaces can be displayed as phase-plane diagrams. The following commands can also be implemented as a MATLAB M-file (see Fig 7.6):

```
≫ clf
≫ subplot(221);
≫ plot(x(:,1),y(:,1));
≫ title('-1 * X2');
≫ xlabel('X1');
≫ ylabel('X2');
```

```
>> subplot(222);
>> plot(x(:,3),y(:,3));
>> title('-0.7 * X2');
>> xlabel('X1');
>> ylabel('X2');

>> subplot(223);
>> plot(x(:,6),y(:,6));
>> title('-0.4 * X2');
>> xlabel('X1');
>> ylabel('X2');

>> subplot(224);
>> plot(x(:,9),y(:,9));
>> title('-0.1 * X2');
>> xlabel('X1');
>> ylabel('X2');

>> subplot
```

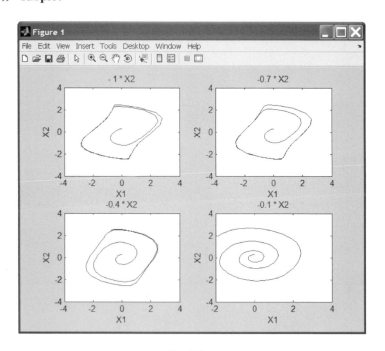

Fig. 7.6

7.4 Conditional System Model

This system provides a demonstration of a simple conditional **if** test implemented as a Stateflow model. All the Stateflow portion of this diagram does is provide the input (**Ramp**) and the output (**Scope**).

The following MATLAB command line loads the system:

≫ **sf_if;**

Figure 7.7 shows the model opened.

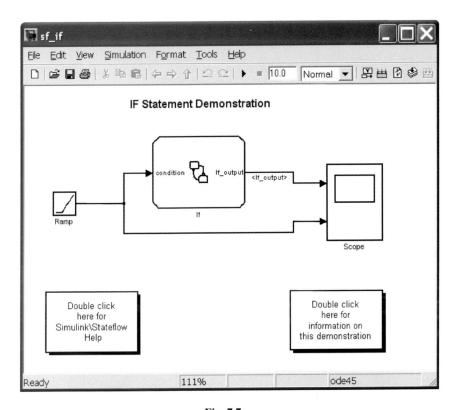

Fig. 7.7

The Stateflow model is equivalent to the following M-file:

```
% if test M-file
if condition= 20
   If_output = condition^2;
elseif condition > 20 & condition < = 50
   If_output = 50*sin(condition)
Else
   If_output = condition^2;
end
%
```

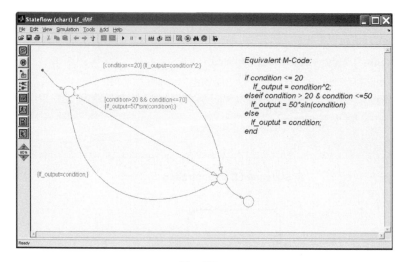

Fig. 7.8

A double-click on the **if** block opens the Stateflow model shown in Fig. 7.8. This model is executed using the following command:

>> **sim('sf_if ',10)**

When the model is executed, the Stateflow model shows the change in state with time. In this case, the condition is a ramp input with a slope of 10 for the first 10 seconds. The resulting time response is shown in Fig. 7.9.

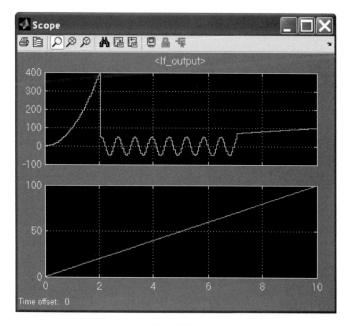

Fig. 7.9

7.5 Combined Simulink® and Stateflow® Systems

This system demonstrates the use of a Stateflow model within Simulink. It also shows how a custom interface can be written by the user to directly enter parameters into the model without the use of the MATLAB **Command Window**. The model is of a forced mass sliding on a surface with friction with spring damping. The model can be loaded using the following command (see Fig. 7.10):

>> **sf_stickslip**

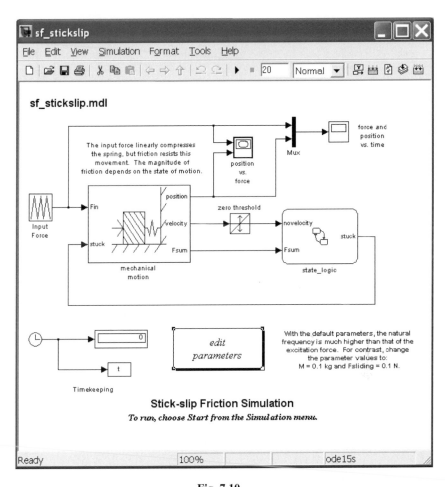

Fig. 7.10

The state logic for this system was programmed in Stateflow software and was stored under the Stateflow model name **state_logic**. This block should have opened when you opened the Simulink model **sf_stickslip**. If not, just double-click on the block in the Simulink window. The **state_logic** window should appear as in Fig. 7.11.

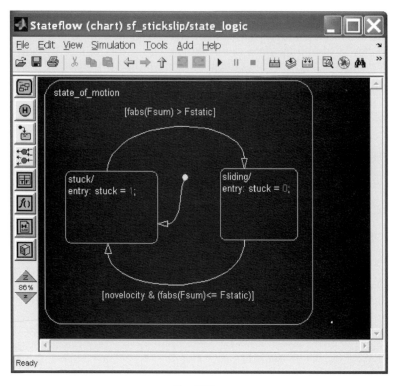

Fig. 7.11

You can now try different model parameters by double-clicking on the Simulink block **edit parameters**. Doing so will open the window shown in Fig. 7.12.

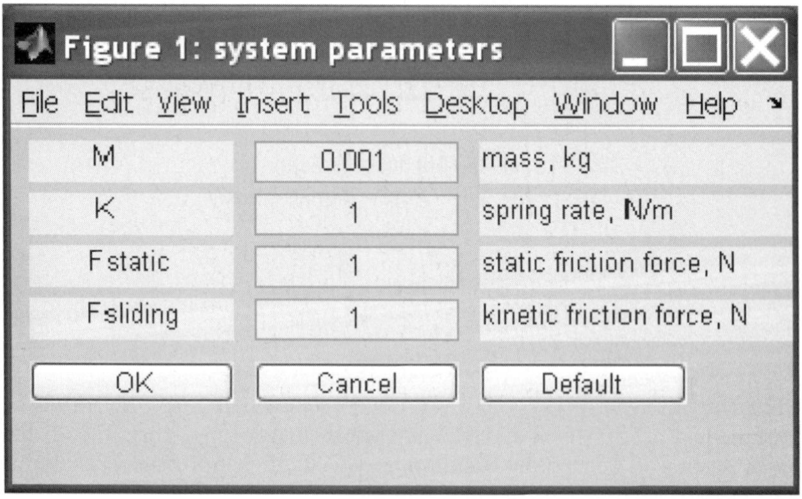

M	0.001	mass, kg
K	1	spring rate, N/m
F static	1	static friction force, N
F sliding	1	kinetic friction force, N

OK   Cancel   Default

Fig. 7.12

Before the model is run, there are no output signals on the **scope** plot (see Fig. 7.13). If the **scope** plot is not open, double-click on the **scope** labeled **force and position vs. time** to open the plot window shown in Fig. 7.10.

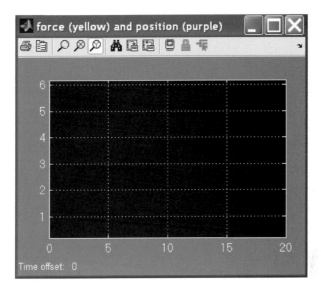

Fig. 7.13

Executing the model using the MATLAB command **sim('sf_stickslip',20)** will give the plots shown in Figs 7.14 and 7.15.

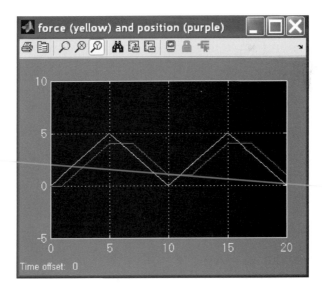

Fig. 7.14

position vs. force
XY Plot
6
5
4
3
2
1
0
-1
Y Axis
0
2
4
6
X Axis

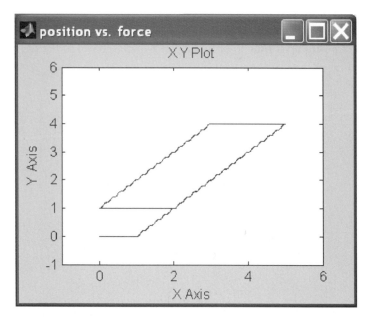

Fig. 7.15

You should have been able to see the changes in the states within the Stateflow model as it executed. Note that the natural frequency of the system is much higher than that of the excitation force. Now change the mass to M = 1 kg and rerun the model. Figure 7.16 shows these modified values.

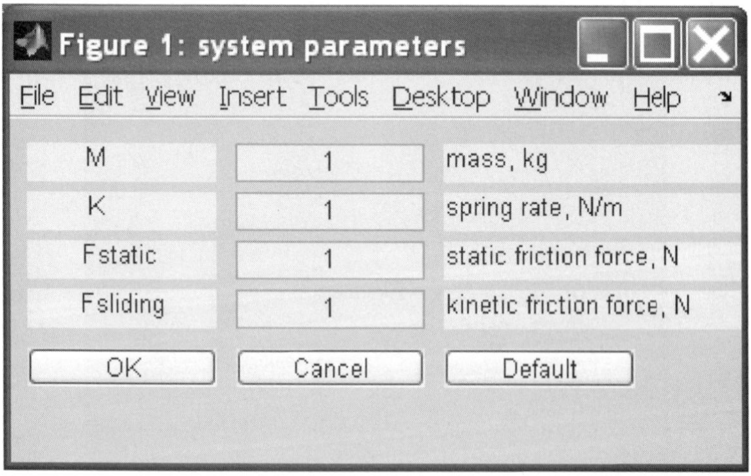

Figure 1: system parameters
File Edit View Insert Tools Desktop Window Help
M
1
mass, kg
K
1
spring rate, N/m
Fstatic
1
static friction force, N
Fsliding
1
kinetic friction force, N
OK
Cancel
Default

Fig. 7.16

The results of this action are shown in Figs 7.17 and 7.18.

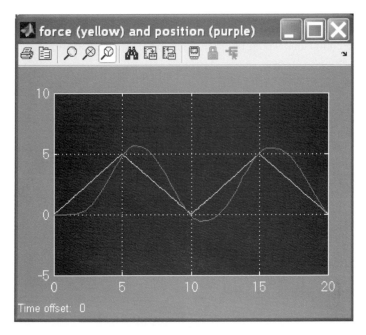

Fig. 7.17

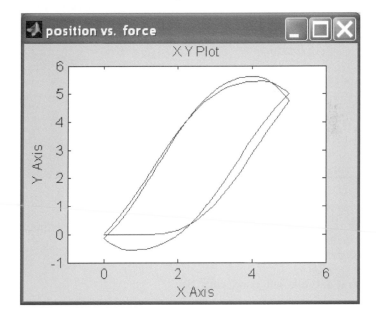

Fig. 7.18

The model can be closed using the following command:

≫ **close_system('sf_stickslip');**

These commands would be used if the model was called from a MATLAB M-file.

7.6 Model Comparison

Compare the implementation of a state-space controller [A,B,C,D] in self-conditioned form versus the standard state-space form. This model requires the Control System Toolbox. It provides a simple demonstration of how different model implementations can be studied using Simulink.

For the self-conditioned state-space controller, if the measured control value is equal to the commanded value (**u_meas = u_dem**), then the controller implementation is the typical state-space controller [A,B,C,D]. If the measured control value (**u_meas**) is limited, then the poles of the controller become those defined in the mask dialog box.

The command to load the system is (see Fig. 7.19)

≫ **aeroblk_self_cond_cntr;**

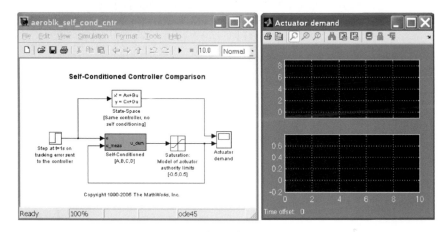

Fig. 7.19

To see the system model inside the two different block descriptions, open them by double-clicking on them.

This system can be simulated from the command line

≫ **sim('aeroblk_self_cond_cntr',10);**

The results of a typical state-space controller [A,B,C,D] and a self-conditioned state-space controller with a limited measured control value are shown in Fig. 7.20.

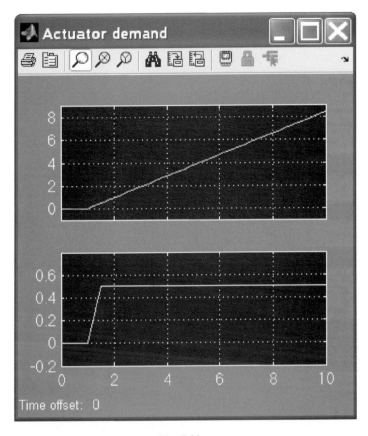

Fig. 7.20

To simulate just the first 2 seconds, rerun the system using the command (see Fig. 7.21)

≫ **sim('aeroblk_self_cond_cntr',2);**

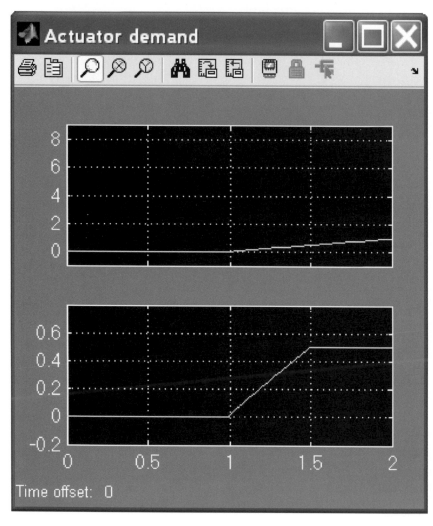

Fig. 7.21

If you click on the **aeroblk_self_cond_cntr** window, you can start and stop the simulation of this system yourself from the **Simulation** menu.

The Simulink model is closed using the command

close_system('aeroblk_self_cond_cntr',0);

7.7 F-14 Control System

This is a model of the one-dimensional (vertical) behavior of an F-14 fighter. It shows how data for a Simulink model can be stored in a separate M-file and loaded as needed for a particular simulation run.

The command to load the data required by the F-14 model is

>> **f14dat;**

This is accomplished by loading and executing the following M-file:

```
% Numerical data for F-14 demo

% Copyright 1990-2002 The MathWorks, Inc.
% $Revision: 1.16 $

g = 32.2;
Uo = 689.4000;
Vto = 690.4000;
% Stability derivatives
Mw = -0.00592;
Mq = -0.6571;
Md = -6.8847;
Zd = -63.9979;
Zw = -0.6385;
% Gains
cmdgain = 3.490954472728077e-02;
Ka = 0.6770;
Kq = 0.8156;
Kf = -1.7460;
Ki = -3.8640;
% Other constants
a = 2.5348;
Gamma = 0.0100;
b = 64.1300;
Beta = 426.4352;
Sa = 0.005236;
Swg = 3;
Ta = 0.0500;
Tal = 0.3959;
Ts = 0.1000;
W1 = 2.9710;
W2 = 4.1440;
Wa = 10;
```

Next load the system and associated numerical constants (see Fig. 7.22):

>> **f14;**

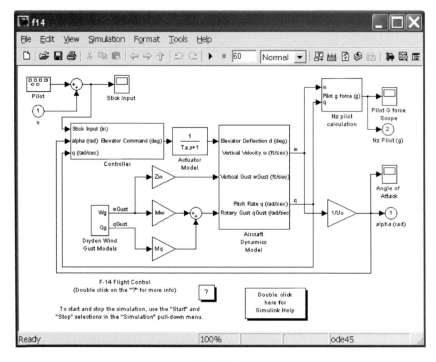

Fig. 7.22

Select the two graphics windows to analyze the system by double-clicking on the two **Scopes Pilot G force Scope** and **Angle of Attack** if they do not open when the F-14 model opens.

The resulting time histories are shown in Figs 7.23 and 7.24.

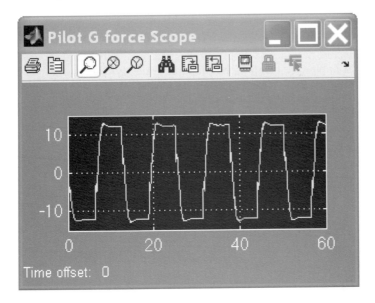

Fig. 7.23

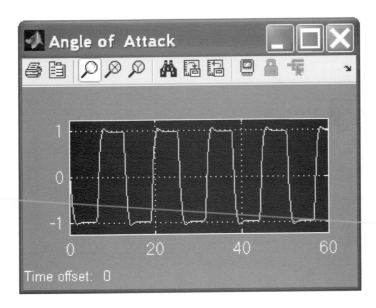

Fig. 7.24

You can now change values in the M-file **f14dat** using the MATLAB M-file Editor.

7.8 Conclusion

This lecture was a very brief introduction to Simulink and its capabilities. Further examples are given in the Simulink manual. Real-time simulations can be run using the Real-Time Workshop® software.

Other toolboxes are available from The MathWorks specifically for use with Simulink.

Practice Exercises

7.1 Execute the existing Simulink model **sldemo_bounce** from the MATLAB **Command Window**.

In this model a rubber ball is thrown into the air with a velocity of 15 meters/ second from a height of 10 meters. The position of the ball is shown in the lower plot of the scope, and the velocity of the ball is shown in the upper plot.

This system uses a resetable integrator to change the direction of the ball as it comes into contact with the ground; the zero-crossing detection prevents the ball from going below the ground.

From the MATLAB **Command Window** (see Fig. 7.25), drop the ball from two different heights (10 meters and 25 meters), and study the effect of different ball elasticities by testing three different ball elasticity values (-0.8, -0.5, -0.2).

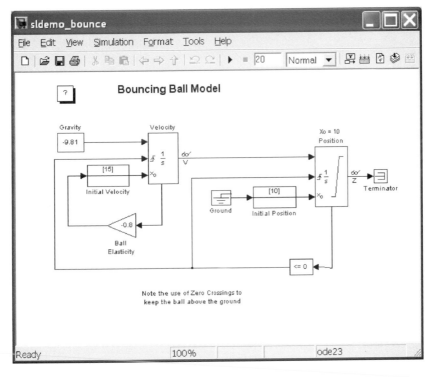

Fig. 7.25

Notes

Introduction to Simulink®

8.1 Introduction and Objectives

This lecture introduces the reader to Simulink® graphical modeling capabilities. It assumes a basic familiarity with MATLAB®.

Upon completion of the second half of this book, the reader will be able to identify the graphics modeling capabilities of Simulink, open and close Simulink models from MATLAB, generate Simulink models, execute and simulate systems implemented in Simulink, modify Simulink model blocks parameters from Simulink and MATLAB, and modify and generate Stateflow® model blocks and Simulink/Stateflow diagrams.

8.2 Standard Simulink® Libraries

Simulink uses block diagrams to represent dynamic systems. Defining a system is much like drawing a block diagram. Blocks are copied from a library of blocks. Connections and other utilities are similarly drawn from block libraries. Special purpose blocks can be purchased in the form of toolboxes from The MathWorks and from third-party vendors. Finally, users can define their own special purpose blocks.

Simulink is started from the main MATLAB window. After the MATLAB program is started the standard main MATLAB interface window appears as shown in Fig. 8.1.

Fig. 8.1

Typing the command **simulink** in the **MATLAB Command Window** opens the standard block library. The standard block library is divided into several subsystems, grouping blocks according to their behavior. Examples using the standard block libraries provided with Simulink are shown in Fig. 8.2. They include sources (which generate input signals), sinks (places where output data can be stored), linear blocks, nonlinear blocks, discrete blocks, and connections. The user can also construct blocks and block libraries.

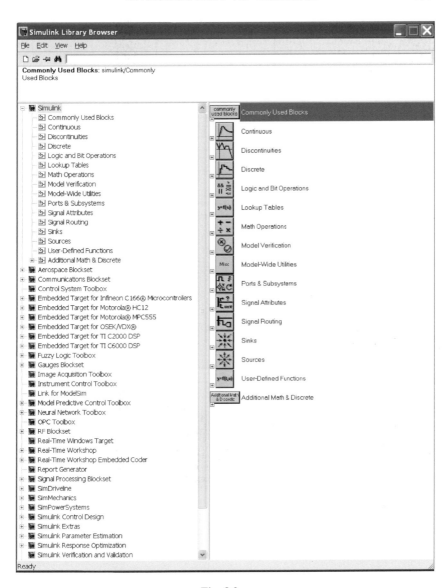

Fig. 8.2

The following are in the **Commonly Used Blocks** Simulink block library. They contain a collection of the most frequently used Simulink blocks from the other libraries.

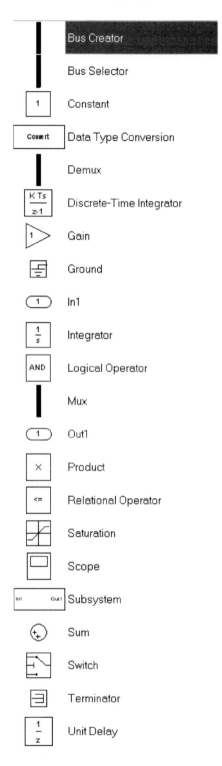

Bus Creator

Bus Selector

Constant

Data Type Conversion

Demux

Discrete-Time Integrator

Gain

Ground

In1

Integrator

Logical Operator

Mux

Out1

Product

Relational Operator

Saturation

Scope

Subsystem

Sum

Switch

Terminator

Unit Delay

The following are in the **Continuous** Simulink block library. They contain continuous dynamic system models.

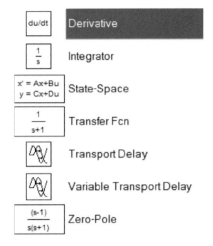

The following are in the **Discontinuities** Simulink block library. They contain several discontinuous nonlinearities.

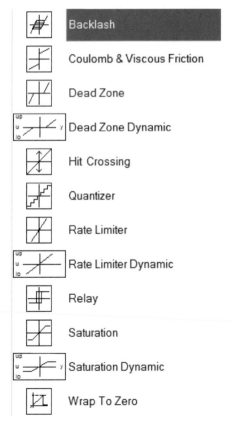

The following are in the **Discrete** Simulink block library. They contain z-transfer functions and time delays.

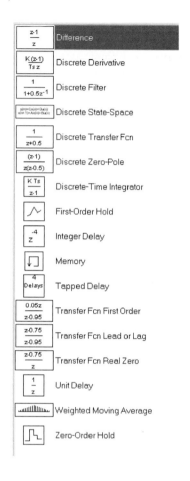

The following are in the main Simulink block library of **Lookup Tables**.

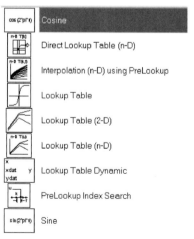

The following are in the Simulink block library of **Logic and Bit Operations**. This library contains a wide variety of bit operators and value and interval tests.

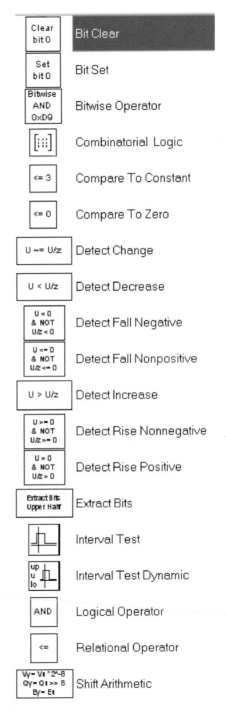

Clear bit 0	Bit Clear
Set bit 0	Bit Set
Bitwise AND 0xD9	Bitwise Operator
[:::]	Combinatorial Logic
<= 3	Compare To Constant
<= 0	Compare To Zero
U ~= U/z	Detect Change
U < U/z	Detect Decrease
U < 0 & NOT U/z < 0	Detect Fall Negative
U <= 0 & NOT U/z <= 0	Detect Fall Nonpositive
U > U/z	Detect Increase
U >= 0 & NOT U/z >= 0	Detect Rise Nonnegative
U > 0 & NOT U/z > 0	Detect Rise Positive
Extract Bits Upper Half	Extract Bits
	Interval Test
up u lo	Interval Test Dynamic
AND	Logical Operator
<=	Relational Operator
Wy= Vi * 2^-8 Qy= Qi >> 8 Ey= Ei	Shift Arithmetic

The following are in the Simulink block library of **Math Operations**.

This library contains a wide variety of functions, assignments, and logic elements.

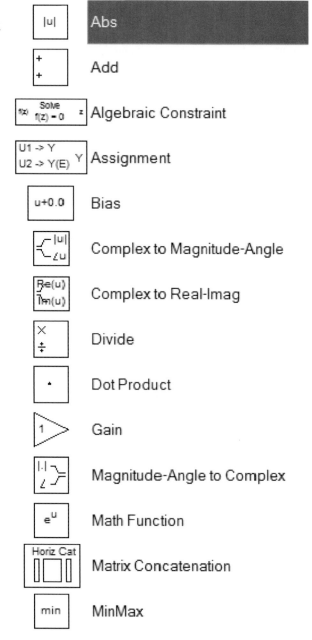

Abs

Add

Algebraic Constraint

Assignment

Bias

Complex to Magnitude-Angle

Complex to Real-Imag

Divide

Dot Product

Gain

Magnitude-Angle to Complex

Math Function

Matrix Concatenation

MinMax

u min(u,y) y / R	MinMax Running Resettable
P(u) O(P) = 5	Polynomial
×	Product
∏	Product of Elements
Re Im	Real-Imag to Complex
U(:)	Reshape
floor	Rounding Function
	Sign
t	Sine Wave Function
1	Slider Gain
+ −	Subtract
(+ +)	Sum
Σ	Sum of Elements
sin	Trigonometric Function
-u	Unary Minus
u+Ts	Weighted Sample Time Math

The following are in the Simulink block library of **Signal Attributes**. These are useful in analyzing the characteristics of the signals generated by your model, including signals internal to the model.

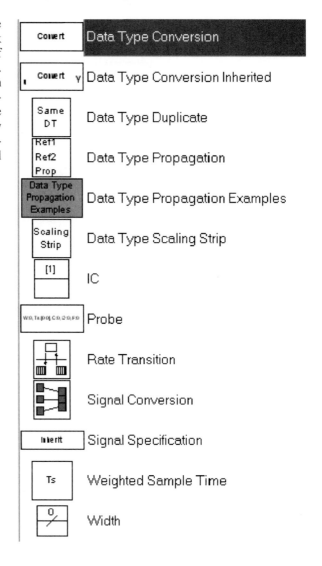

The following are in the Simulink **Model Verification** block library.

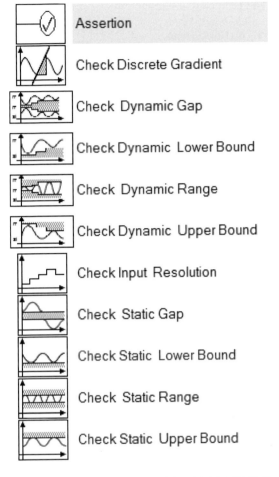

Assertion

Check Discrete Gradient

Check Dynamic Gap

Check Dynamic Lower Bound

Check Dynamic Range

Check Dynamic Upper Bound

Check Input Resolution

Check Static Gap

Check Static Lower Bound

Check Static Range

Check Static Upper Bound

The following are in the **Model-Wide Utilities** Simulink block library.

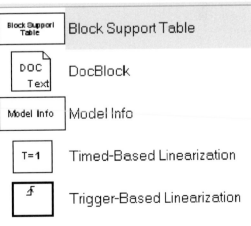

Block Support Table

DocBlock

Model Info

Timed-Based Linearization

Trigger-Based Linearization

The following are in the Simulink block library of **Ports & Subsystems**.

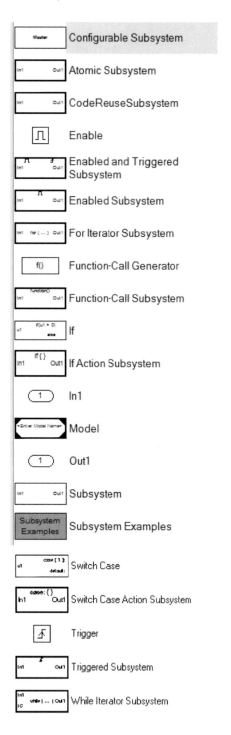

The following are Simulink block library **Signal Routing** elements. Included here are several bus-related models. Several data storage and recall mechanizations are included. Go to and from functions are also available.

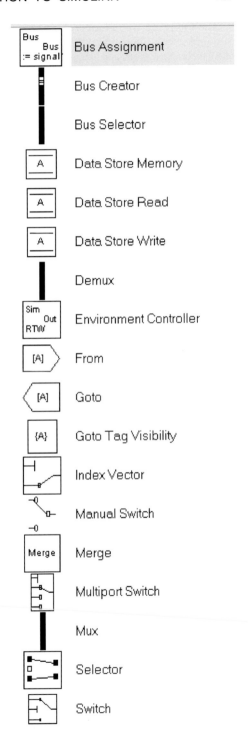

Bus Assignment

Bus Creator

Bus Selector

Data Store Memory

Data Store Read

Data Store Write

Demux

Environment Controller

From

Goto

Goto Tag Visibility

Index Vector

Manual Switch

Merge

Multiport Switch

Mux

Selector

Switch

The following are in the **Sinks** Simulink block library. **Sinks** are used to display plots and data as the Simulink model is executed. **Scope** blocks are the most common way of displaying plots of signal data as it is being generated. Data can also be directly output into MATLAB using the **To File** block. This makes it possible to do further processing and analysis within the MATLAB **Command Window**.

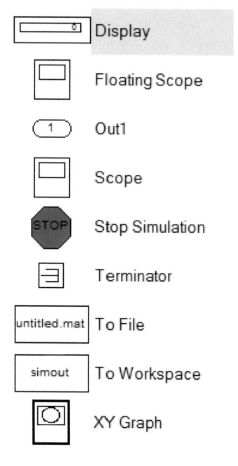

The following are in the **Sources** Simulink block library. Sources provide a wide variety of input signal types. Common input signals for system testing and analysis, such as steps, sine waves, pulses, and random numbers, are provided here. Users can also generate their own signals or input signal data from external sources.

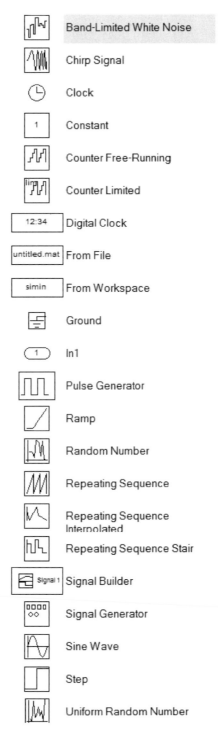

Band-Limited White Noise

Chirp Signal

Clock

Constant

Counter Free-Running

Counter Limited

Digital Clock

From File

From Workspace

Ground

In1

Pulse Generator

Ramp

Random Number

Repeating Sequence

Repeating Sequence Interpolated

Repeating Sequence Stair

Signal Builder

Signal Generator

Sine Wave

Step

Uniform Random Number

The following are in the Simulink block library of **User Defined Functions**. The **S-Function** is equivalent to the M-file written in the MATLAB command language, or the MEX-function written in C, C++, FORTRAN, and/or Ada. Although M-files and MEX functions are executed in the MATLAB **Command Window**, **S-Functions** are directly included within the Simulink model. An **S-Function Builder** is included to aid the user in constructing an **S-Function**. Note that MATLAB functions can be directly incorporated within the Simulink model using the **MATLAB Fcn** block.

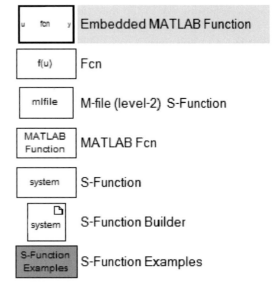

The following are in the **Additional Math & Discrete** Simulink block library.

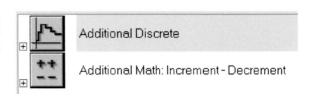

8.3 Simulink® Aerospace Blockset

Figure 8.3 represents all of the Aerospace Block Libraries contained within the Simulink **Aerospace Blockset**. Many different aerospace modeling utilities, which in the past needed to be constructed as Simulink models or **S-Functions**, are now provided by The MathWorks in this set of libraries. The models contained within this blockset are described within the following sections.

Fig. 8.3

The following are the **Actuators** in the Simulink **Aerospace Blockset** library. Simple linear and nonlinear models are available.

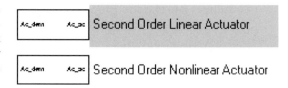

The following are in the Simulink **Aerospace Aerodynamics** library.

The following are in the Simulink **Aerospace Blockset Animation** library.

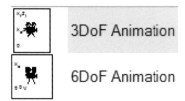

The following are in the Simulink **Aerospace Blockset Environment** library.

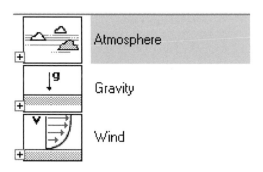

The following are the 3- and 6-DOF **Equations of Motion** blocks in the Simulink **Aerospace Blockset** library.

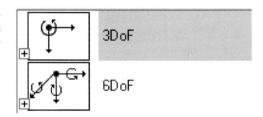

The following are in the Simulink **Aerospace Blockset Flight Parameters** library. Calculate the angles between the body and the velocity vector (incidence and sideslip) and the velocity magnitude from the components in body axes (V_b).

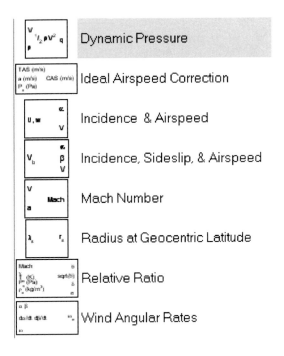

The following are in the Simulink **Aerospace Blockset GNC** library. A large variety of linear state-space forms are provided in this library for system representations.

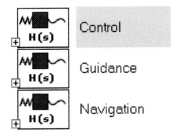

The following are in the Simulink **Aerospace Blockset Mass Properties** library. One block is used to calculate the center of gravity location. Linear interpolation is also used to determine center of gravity as a function of mass.

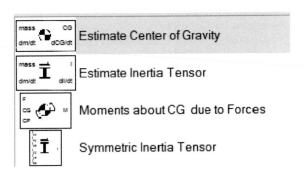

The following simple turbofan engine model is in the Simulink **Aerospace Blockset Propulsion** library.

The following are the **Transformations** in the Simulink **Aerospace Utilities Blockset** library.

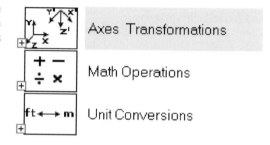

8.4   Simulink® Installation and Demonstrations

The following is the complete MATLAB/Simulink path. It is given for a full MATLAB/Simulink/Stateflow installation of Release 2006b.

The Simulink toolboxes and other utilities you have available in your MATLAB installation can be accessed using the command (this is the same as the command for MATLAB toolboxes, etc.).

≫ **path**

```
C:\Program Files\MATLAB\R2006b\toolbox\matlab\general
C:\Program Files\MATLAB\R2006b\toolbox\matlab\ops
C:\Program Files\MATLAB\R2006b\toolbox\matlab\lang
C:\Program Files\MATLAB\R2006b\toolbox\matlab\elmat
C:\Program Files\MATLAB\R2006b\toolbox\matlab\elfun
C:\Program Files\MATLAB\R2006b\toolbox\matlab\specfun
C:\Program Files\MATLAB\R2006b\toolbox\matlab\matfun
C:\Program Files\MATLAB\R2006b\toolbox\matlab\datafun
C:\Program Files\MATLAB\R2006b\toolbox\matlab\polyfun
C:\Program Files\MATLAB\R2006b\toolbox\matlab\funfun
C:\Program Files\MATLAB\R2006b\toolbox\matlab\sparfun
C:\Program Files\MATLAB\R2006b\toolbox\matlab\scribe
C:\Program Files\MATLAB\R2006b\toolbox\matlab\graph2d
C:\Program Files\MATLAB\R2006b\toolbox\matlab\graph3d
C:\Program Files\MATLAB\R2006b\toolbox\matlab\specgraph
C:\Program Files\MATLAB\R2006b\toolbox\matlab\graphics
C:\Program Files\MATLAB\R2006b\toolbox\matlab\uitools
C:\Program Files\MATLAB\R2006b\toolbox\matlab\strfun
C:\Program Files\MATLAB\R2006b\toolbox\matlab\imagesci
```

C:\Program Files\MATLAB\R2006b\toolbox\matlab\iofun
C:\Program Files\MATLAB\R2006b\toolbox\matlab\audiovideo
C:\Program Files\MATLAB\R2006b\toolbox\matlab\timefun
C:\Program Files\MATLAB\R2006b\toolbox\matlab\datatypes
C:\Program Files\MATLAB\R2006b\toolbox\matlab\verctrl
C:\Program Files\MATLAB\R2006b\toolbox\matlab\codetools
C:\Program Files\MATLAB\R2006b\toolbox\matlab\helptools
C:\Program Files\MATLAB\R2006b\toolbox\matlab\winfun
C:\Program Files\MATLAB\R2006b\toolbox\matlab\demos
C:\Program Files\MATLAB\R2006b\toolbox\matlab\timeseries
C:\Program Files\MATLAB\R2006b\toolbox\matlab\hds
C:\Program Files\MATLAB\R2006b\toolbox\local
C:\Program Files\MATLAB\R2006b\toolbox\shared\controllib
C:\Program Files\MATLAB\R2006b\toolbox\simulink\simulink
C:\Program Files\MATLAB\R2006b\toolbox\simulink\blocks
C:\Program Files\MATLAB\R2006b\toolbox\simulink\components
C:\Program Files\MATLAB\R2006b\toolbox\simulink\fixedandfloat
C:\Program Files\MATLAB\R2006b\toolbox\simulink\fixedandfloat\fxpdemos
C:\Program Files\MATLAB\R2006b\toolbox\simulink\fixedandfloat\obsolete
C:\Program Files\MATLAB\R2006b\toolbox\simulink\simdemos
C:\Program Files\MATLAB\R2006b\toolbox\simulink\simdemos\aerospace
C:\Program Files\MATLAB\R2006b\toolbox\simulink\simdemos\automotive
C:\Program Files\MATLAB\R2006b\toolbox\simulink\simdemos\simfeatures
C:\Program Files\MATLAB\R2006b\toolbox\simulink\simdemos\simgeneral
C:\Program Files\MATLAB\R2006b\toolbox\simulink\dee
C:\Program Files\MATLAB\R2006b\toolbox\shared\dastudio
C:\Program Files\MATLAB\R2006b\toolbox\stateflow\stateflow
C:\Program Files\MATLAB\R2006b\toolbox\rtw\rtw
C:\Program Files\MATLAB\R2006b\toolbox\simulink\simulink\modeladvisor
C:\Program Files\MATLAB\R2006b\toolbox\simulink\simulink\
modeladvisor\fixpt
C:\Program Files\MATLAB\R2006b\toolbox\simulink\simulink\MPlayIO
C:\Program Files\MATLAB\R2006b\toolbox\simulink\simulink\dataobjectwizard
C:\Program Files\MATLAB\R2006b\toolbox\shared\fixedpointlib
C:\Program Files\MATLAB\R2006b\toolbox\stateflow\sfdemos
C:\Program Files\MATLAB\R2006b\toolbox\stateflow\coder
C:\Program Files\MATLAB\R2006b\toolbox\rtw\rtwdemos
C:\Program Files\MATLAB\R2006b\toolbox\rtw\rtwdemos\rsimdemos
C:\Program Files\MATLAB\R2006b\toolbox\rtw\targets\asap2\asap2
C:\Program Files\MATLAB\R2006b\toolbox\rtw\targets\asap2\asap2\user
C:\Program Files\MATLAB\R2006b\toolbox\rtw\targets\common\can\blocks
C:\Program Files\MATLAB\R2006b\toolbox\rtw\targets\common\configuration\
resource
C:\Program Files\MATLAB\R2006b\toolbox\rtw\targets\common\tgtcommon
C:\Program Files\MATLAB\R2006b\toolbox\rtw\targets\rtwin\rtwin
C:\Program Files\MATLAB\R2006b\toolbox\simulink\accelerator
C:\Program Files\MATLAB\R2006b\toolbox\simulink\accelerator\acceldemos
C:\Program Files\MATLAB\R2006b\toolbox\rtw\accel

C:\Program Files\MATLAB\R2006b\toolbox\aeroblks\aeroblks
C:\Program Files\MATLAB\R2006b\toolbox\aeroblks\aerodemos
C:\Program Files\MATLAB\R2006b\toolbox\aeroblks\aerodemos\texture
C:\Program Files\MATLAB\R2006b\toolbox\bioinfo\bioinfo
C:\Program Files\MATLAB\R2006b\toolbox\bioinfo\biolearning
C:\Program Files\MATLAB\R2006b\toolbox\bioinfo\microarray
C:\Program Files\MATLAB\R2006b\toolbox\bioinfo\mass_spec
C:\Program Files\MATLAB\R2006b\toolbox\bioinfo\proteins
C:\Program Files\MATLAB\R2006b\toolbox\bioinfo\biomatrices
C:\Program Files\MATLAB\R2006b\toolbox\bioinfo\biodemos
C:\Program Files\MATLAB\R2006b\toolbox\rtw\targets\c166\c166
C:\Program Files\MATLAB\R2006b\toolbox\rtw\targets\c166\blocks
C:\Program Files\MATLAB\R2006b\toolbox\rtw\targets\c166\c166demos
C:\Program Files\MATLAB\R2006b\toolbox\ccslink\ccslink
C:\Program Files\MATLAB\R2006b\toolbox\ccslink\ccslink\outproc
C:\Program Files\MATLAB\R2006b\toolbox\ccslink\ccsblks
C:\Program Files\MATLAB\R2006b\toolbox\ccslink\ccsdemos
C:\Program Files\MATLAB\R2006b\toolbox\comm\comm
C:\Program Files\MATLAB\R2006b\toolbox\comm\commdemos
C:\Program Files\MATLAB\R2006b\toolbox\comm\commdemos\
   commdocdemos
C:\Program Files\MATLAB\R2006b\toolbox\comm\commobsolete
C:\Program Files\MATLAB\R2006b\toolbox\commblks\commblks
C:\Program Files\MATLAB\R2006b\toolbox\commblks\commmasks
C:\Program Files\MATLAB\R2006b\toolbox\commblks\commmex
C:\Program Files\MATLAB\R2006b\toolbox\commblks\commblksdemos
C:\Program Files\MATLAB\R2006b\toolbox\commblks\commblksobsolete\v3
C:\Program Files\MATLAB\R2006b\toolbox\commblks\commblksobsolete\v2p5
C:\Program Files\MATLAB\R2006b\toolbox\commblks\commblksobsolete\v2
C:\Program Files\MATLAB\R2006b\toolbox\commblks\commblksobsolete\v1p5
C:\Program Files\MATLAB\R2006b\toolbox\control\control
C:\Program Files\MATLAB\R2006b\toolbox\control\ctrlguis
C:\Program Files\MATLAB\R2006b\toolbox\control\ctrlobsolete
C:\Program Files\MATLAB\R2006b\toolbox\control\ctrlutil
C:\Program Files\MATLAB\R2006b\toolbox\control\ctrldemos
C:\Program Files\MATLAB\R2006b\toolbox\curvefit\curvefit
C:\Program Files\MATLAB\R2006b\toolbox\curvefit\cftoolgui
C:\Program Files\MATLAB\R2006b\toolbox\shared\optimlib
C:\Program Files\MATLAB\R2006b\toolbox\daq\daq
C:\Program Files\MATLAB\R2006b\toolbox\daq\daqguis
C:\Program Files\MATLAB\R2006b\toolbox\daq\daqdemos
C:\Program Files\MATLAB\R2006b\toolbox\database\database
C:\Program Files\MATLAB\R2006b\toolbox\database\dbdemos
C:\Program Files\MATLAB\R2006b\toolbox\database\vqb
C:\Program Files\MATLAB\R2006b\toolbox\datafeed\datafeed
C:\Program Files\MATLAB\R2006b\toolbox\datafeed\dfgui
C:\Program Files\MATLAB\R2006b\toolbox\des\desblks
C:\Program Files\MATLAB\R2006b\toolbox\des\desmasks

C:\Program Files\MATLAB\R2006b\toolbox\des\desmex
C:\Program Files\MATLAB\R2006b\toolbox\des\desdemos
C:\Program Files\MATLAB\R2006b\toolbox\physmod\drive\drive
C:\Program Files\MATLAB\R2006b\toolbox\physmod\drive\drivedemos
C:\Program Files\MATLAB\R2006b\toolbox\dspblks\dspblks
C:\Program Files\MATLAB\R2006b\toolbox\dspblks\dspmasks
C:\Program Files\MATLAB\R2006b\toolbox\dspblks\dspmex
C:\Program Files\MATLAB\R2006b\toolbox\dspblks\dspdemos
C:\Program Files\MATLAB\R2006b\toolbox\rtw\targets\ecoder
C:\Program Files\MATLAB\R2006b\toolbox\rtw\targets\ecoder\ecoderdemos
C:\Program Files\MATLAB\R2006b\toolbox\rtw\targets\mpt
C:\Program Files\MATLAB\R2006b\toolbox\rtw\targets\mpt\mpt
C:\Program Files\MATLAB\R2006b\toolbox\rtw\targets\mpt\user_specific
C:\Program Files\MATLAB\R2006b\toolbox\exlink
C:\Program Files\MATLAB\R2006b\toolbox\symbolic\extended
C:\Program Files\MATLAB\R2006b\toolbox\filterdesign\filterdesign
C:\Program Files\MATLAB\R2006b\toolbox\filterdesign\quantization
C:\Program Files\MATLAB\R2006b\toolbox\filterdesign\filtdesdemos
C:\Program Files\MATLAB\R2006b\toolbox\finance\finance
C:\Program Files\MATLAB\R2006b\toolbox\finance\calendar
C:\Program Files\MATLAB\R2006b\toolbox\finance\findemos
C:\Program Files\MATLAB\R2006b\toolbox\finance\finsupport
C:\Program Files\MATLAB\R2006b\toolbox\finance\ftseries
C:\Program Files\MATLAB\R2006b\toolbox\finance\ftsdemos
C:\Program Files\MATLAB\R2006b\toolbox\finance\ftsdata
C:\Program Files\MATLAB\R2006b\toolbox\finance\ftstutorials
C:\Program Files\MATLAB\R2006b\toolbox\finderiv\finderiv
C:\Program Files\MATLAB\R2006b\toolbox\finfixed\finfixed
C:\Program Files\MATLAB\R2006b\toolbox\fixedpoint\fixedpoint
C:\Program Files\MATLAB\R2006b\toolbox\fixedpoint\fidemos
C:\Program Files\MATLAB\R2006b\toolbox\fixedpoint\fimex
C:\Program Files\MATLAB\R2006b\toolbox\fixpoint
C:\Program Files\MATLAB\R2006b\toolbox\fuzzy\fuzzy
C:\Program Files\MATLAB\R2006b\toolbox\fuzzy\fuzdemos
C:\Program Files\MATLAB\R2006b\toolbox\gads
C:\Program Files\MATLAB\R2006b\toolbox\gads\gads
C:\Program Files\MATLAB\R2006b\toolbox\gads\gadsdemos
C:\Program Files\MATLAB\R2006b\toolbox\garch\garch
C:\Program Files\MATLAB\R2006b\toolbox\garch\garchdemos
C:\Program Files\MATLAB\R2006b\toolbox\gauges
C:\Program Files\MATLAB\R2006b\toolbox\rtw\targets\hc12\hc12
C:\Program Files\MATLAB\R2006b\toolbox\rtw\targets\hc12\blocks
C:\Program Files\MATLAB\R2006b\toolbox\rtw\targets\hc12\codewarrior
C:\Program Files\MATLAB\R2006b\toolbox\rtw\targets\hc12\hc12demos
C:\Program Files\MATLAB\R2006b\toolbox\hdlfilter\hdlfilter
C:\Program Files\MATLAB\R2006b\toolbox\hdlfilter\hdlfiltdemos
C:\Program Files\MATLAB\R2006b\toolbox\shared\hdlshared
C:\Program Files\MATLAB\R2006b\toolbox\ident\ident

C:\Program Files\MATLAB\R2006b\toolbox\ident\idobsolete
C:\Program Files\MATLAB\R2006b\toolbox\ident\idguis
C:\Program Files\MATLAB\R2006b\toolbox\ident\idutils
C:\Program Files\MATLAB\R2006b\toolbox\ident\iddemos
C:\Program Files\MATLAB\R2006b\toolbox\ident\idhelp
C:\Program Files\MATLAB\R2006b\toolbox\images\images
C:\Program Files\MATLAB\R2006b\toolbox\images\imuitools
C:\Program Files\MATLAB\R2006b\toolbox\images\imdemos
C:\Program Files\MATLAB\R2006b\toolbox\images\iptutils
C:\Program Files\MATLAB\R2006b\toolbox\shared\imageslib
C:\Program Files\MATLAB\R2006b\toolbox\images\medformats
C:\Program Files\MATLAB\R2006b\toolbox\imaq\imaq
C:\Program Files\MATLAB\R2006b\toolbox\shared\imaqlib
C:\Program Files\MATLAB\R2006b\toolbox\imaq\imaqdemos
C:\Program Files\MATLAB\R2006b\toolbox\imaq\imaqblks\imaqblks
C:\Program Files\MATLAB\R2006b\toolbox\imaq\imaqblks\imaqmasks
C:\Program Files\MATLAB\R2006b\toolbox\imaq\imaqblks\imaqmex
C:\Program Files\MATLAB\R2006b\toolbox\instrument\instrument
C:\Program Files\MATLAB\R2006b\toolbox\instrument\instrumentdemos
C:\Program Files\MATLAB\R2006b\toolbox\instrument\instrumentblks\
    instrumentblks
C:\Program Files\MATLAB\R2006b\toolbox\instrument\instrumentblks\
    instrumentmex
C:\Program Files\MATLAB\R2006b\toolbox\map\map
C:\Program Files\MATLAB\R2006b\toolbox\map\mapdemos
C:\Program Files\MATLAB\R2006b\toolbox\map\mapdisp
C:\Program Files\MATLAB\R2006b\toolbox\map\mapformats
C:\Program Files\MATLAB\R2006b\toolbox\map\mapproj
C:\Program Files\MATLAB\R2006b\toolbox\shared\mapgeodegy
C:\Program Files\MATLAB\R2006b\toolbox\mbc\mbc
C:\Program Files\MATLAB\R2006b\toolbox\mbc\mbcdata
C:\Program Files\MATLAB\R2006b\toolbox\mbc\mbcdesign
C:\Program Files\MATLAB\R2006b\toolbox\mbc\mbcexpr
C:\Program Files\MATLAB\R2006b\toolbox\mbc\mbcguitools
C:\Program Files\MATLAB\R2006b\toolbox\mbc\mbclayouts
C:\Program Files\MATLAB\R2006b\toolbox\mbc\mbcmodels
C:\Program Files\MATLAB\R2006b\toolbox\mbc\mbcsimulink
C:\Program Files\MATLAB\R2006b\toolbox\mbc\mbctools
C:\Program Files\MATLAB\R2006b\toolbox\mbc\mbcview
C:\Program Files\MATLAB\R2006b\toolbox\physmod\mech\mech
C:\Program Files\MATLAB\R2006b\toolbox\physmod\mech\mechdemos
C:\Program Files\MATLAB\R2006b\toolbox\physmod\pmimport\pmimport
C:\Program Files\MATLAB\R2006b\toolbox\slvnv\simcoverage
C:\Program Files\MATLAB\R2006b\toolbox\modelsim\modelsim
C:\Program Files\MATLAB\R2006b\toolbox\modelsim\modelsimdemos
C:\Program Files\MATLAB\R2006b\toolbox\mpc\mpc
C:\Program Files\MATLAB\R2006b\toolbox\mpc\mpcdemos
C:\Program Files\MATLAB\R2006b\toolbox\mpc\mpcguis

C:\Program Files\MATLAB\R2006b\toolbox\mpc\mpcobsolete
C:\Program Files\MATLAB\R2006b\toolbox\mpc\mpcutils
C:\Program Files\MATLAB\R2006b\toolbox\rtw\targets\mpc555dk
C:\Program Files\MATLAB\R2006b\toolbox\rtw\targets\mpc555dk\
common\configuration
C:\Program Files\MATLAB\R2006b\toolbox\rtw\targets\mpc555dk\
mpc555demos
C:\Program Files\MATLAB\R2006b\toolbox\rtw\targets\mpc555dk\mpc555dk
C:\Program Files\MATLAB\R2006b\toolbox\rtw\targets\mpc555dk\pil
C:\Program Files\MATLAB\R2006b\toolbox\rtw\targets\mpc555dk\rt\
blockset\mfiles
C:\Program Files\MATLAB\R2006b\toolbox\rtw\targets\mpc555dk\rt\blockset
C:\Program Files\MATLAB\R2006b\toolbox\nnet
C:\Program Files\MATLAB\R2006b\toolbox\nnet\nncontrol
C:\Program Files\MATLAB\R2006b\toolbox\nnet\nndemos
C:\Program Files\MATLAB\R2006b\toolbox\nnet\nnet
C:\Program Files\MATLAB\R2006b\toolbox\nnet\nnet\nnanalyze
C:\Program Files\MATLAB\R2006b\toolbox\nnet\nnet\nncustom
C:\Program Files\MATLAB\R2006b\toolbox\nnet\nnet\nndistance
C:\Program Files\MATLAB\R2006b\toolbox\nnet\nnet\nnformat
C:\Program Files\MATLAB\R2006b\toolbox\nnet\nnet\nninit
C:\Program Files\MATLAB\R2006b\toolbox\nnet\nnet\nnlearn
C:\Program Files\MATLAB\R2006b\toolbox\nnet\nnet\nnnetinput
C:\Program Files\MATLAB\R2006b\toolbox\nnet\nnet\nnnetwork
C:\Program Files\MATLAB\R2006b\toolbox\nnet\nnet\nonperformance
C:\Program Files\MATLAB\R2006b\toolbox\nnet\nnet\nnplot
C:\Program Files\MATLAB\R2006b\toolbox\nnet\nnet\nnprocess
C:\Program Files\MATLAB\R2006b\toolbox\nnet\nnet\nnsearch
C:\Program Files\MATLAB\R2006b\toolbox\nnet\nnet\nntopology
C:\Program Files\MATLAB\R2006b\toolbox\nnet\nnet\nntrain
C:\Program Files\MATLAB\R2006b\toolbox\nnet\nnet\nntransfer
C:\Program Files\MATLAB\R2006b\toolbox\nnet\nnet\nnweight
C:\Program Files\MATLAB\R2006b\toolbox\nnet\nnguis
C:\Program Files\MATLAB\R2006b\toolbox\nnet\nnguis\nftool
C:\Program Files\MATLAB\R2006b\toolbox\nnet\nnguis\nntool
C:\Program Files\MATLAB\R2006b\toolbox\nnet\nnobsolete
C:\Program Files\MATLAB\R2006b\toolbox\nnet\nnresource
C:\Program Files\MATLAB\R2006b\toolbox\nnet\nnutils
C:\Program Files\MATLAB\R2006b\toolbox\opc\opc
C:\Program Files\MATLAB\R2006b\toolbox\opc\opcgui
C:\Program Files\MATLAB\R2006b\toolbox\opc\opcdemos
C:\Program Files\MATLAB\R2006b\toolbox\opc\opcdemos\opcblksdemos
C:\Program Files\MATLAB\R2006b\toolbox\opc\opcblks\opcblks
C:\Program Files\MATLAB\R2006b\toolbox\opc\opcblks\opcmasks
C:\Program Files\MATLAB\R2006b\toolbox\optim
C:\Program Files\MATLAB\R2006b\toolbox\rtw\targets\osek\osek
C:\Program Files\MATLAB\R2006b\toolbox\rtw\targets\osek\osekdemos
C:\Program Files\MATLAB\R2006b\toolbox\rtw\targets\osek\blocks

C:\Program Files\MATLAB\R2006b\toolbox\rtw\targets\osek\osekworks
C:\Program Files\MATLAB\R2006b\toolbox\rtw\targets\osek\proosek
C:\Program Files\MATLAB\R2006b\toolbox\pde
C:\Program Files\MATLAB\R2006b\toolbox\physmod\pm_util\pm_util
C:\Program Files\MATLAB\R2006b\toolbox\physmod\powersys\powersys
C:\Program Files\MATLAB\R2006b\toolbox\physmod\powersys\powerdemo
C:\Program Files\MATLAB\R2006b\toolbox\physmod\powersys\drives\drives
C:\Program Files\MATLAB\R2006b\toolbox\physmod\powersys\drives\
    drivesdemo
C:\Program Files\MATLAB\R2006b\toolbox\physmod\powersys\facts\facts
C:\Program Files\MATLAB\R2006b\toolbox\physmod\powersys\
    facts\factsdemo
C:\Program Files\MATLAB\R2006b\toolbox\physmod\powersys\DR\DR
C:\Program Files\MATLAB\R2006b\toolbox\physmod\powersys\DR\DRdemo
C:\Program Files\MATLAB\R2006b\toolbox\slvnv\reqmgt
C:\Program Files\MATLAB\R2006b\toolbox\slvnv\rmidemos
C:\Program Files\MATLAB\R2006b\toolbox\rf\rf
C:\Program Files\MATLAB\R2006b\toolbox\rf\rfdemos
C:\Program Files\MATLAB\R2006b\toolbox\rf\rftool
C:\Program Files\MATLAB\R2006b\toolbox\rfblks\rfblks
C:\Program Files\MATLAB\R2006b\toolbox\rfblks\rfblksmasks
C:\Program Files\MATLAB\R2006b\toolbox\rfblks\rfblksmex
C:\Program Files\MATLAB\R2006b\toolbox\rfblks\rfblksdemos
C:\Program Files\MATLAB\R2006b\toolbox\robust\robust
C:\Program Files\MATLAB\R2006b\toolbox\robust\rctlmi
C:\Program Files\MATLAB\R2006b\toolbox\robust\rctutil
C:\Program Files\MATLAB\R2006b\toolbox\robust\rctdemos
C:\Program Files\MATLAB\R2006b\toolbox\robust\rctobsolete\robust
C:\Program Files\MATLAB\R2006b\toolbox\robust\rctobsolete\lmi
C:\Program Files\MATLAB\R2006b\toolbox\robust\rctobsolete\mutools\
    commands
C:\Program Files\MATLAB\R2006b\toolbox\robust\rctobsolete\mutools\subs
C:\Program Files\MATLAB\R2006b\toolbox\rptgen\rptgen
C:\Program Files\MATLAB\R2006b\toolbox\rptgen\rptgendemos
C:\Program Files\MATLAB\R2006b\toolbox\rptgen\rptgenv1
C:\Program Files\MATLAB\R2006b\toolbox\rptgenext\rptgenext
C:\Program Files\MATLAB\R2006b\toolbox\rptgenext\rptgenextdemos
C:\Program Files\MATLAB\R2006b\toolbox\rptgenext\rptgenextv1
C:\Program Files\MATLAB\R2006b\toolbox\signal\signal
C:\Program Files\MATLAB\R2006b\toolbox\signal\sigtools
C:\Program Files\MATLAB\R2006b\toolbox\signal\sptoolgui
C:\Program Files\MATLAB\R2006b\toolbox\signal\sigdemos
C:\Program Files\MATLAB\R2006b\toolbox\simbio\simbio
C:\Program Files\MATLAB\R2006b\toolbox\simbio\simbiodemos
C:\Program Files\MATLAB\R2006b\toolbox\slcontrol\slcontrol
C:\Program Files\MATLAB\R2006b\toolbox\slcontrol\slctrlguis
C:\Program Files\MATLAB\R2006b\toolbox\slcontrol\slctrlutil

C:\Program Files\MATLAB\R2006b\toolbox\slcontrol\slctrldemos
C:\Program Files\MATLAB\R2006b\toolbox\slestim\slestdemos
C:\Program Files\MATLAB\R2006b\toolbox\slestim\slestguis
C:\Program Files\MATLAB\R2006b\toolbox\slestim\slestim
C:\Program Files\MATLAB\R2006b\toolbox\slestim\slestmex
C:\Program Files\MATLAB\R2006b\toolbox\slestim\slestutil
C:\Program Files\MATLAB\R2006b\toolbox\sloptim\sloptim
C:\Program Files\MATLAB\R2006b\toolbox\sloptim\sloptguis
C:\Program Files\MATLAB\R2006b\toolbox\sloptim\sloptdemos
C:\Program Files\MATLAB\R2006b\toolbox\sloptim\sloptobsolete
C:\Program Files\MATLAB\R2006b\toolbox\slvnv\slvnv
C:\Program Files\MATLAB\R2006b\toolbox\slvnv\simcovdemos
C:\Program Files\MATLAB\R2006b\toolbox\splines
C:\Program Files\MATLAB\R2006b\toolbox\stats
C:\Program Files\MATLAB\R2006b\toolbox\symbolic
C:\Program Files\MATLAB\R2006b\toolbox\rtw\targets\tic2000\tic2000
C:\Program Files\MATLAB\R2006b\toolbox\rtw\targets\tic2000\tic2000blks
C:\Program Files\MATLAB\R2006b\toolbox\rtw\targets\tic2000\tic2000demos
C:\Program Files\MATLAB\R2006b\toolbox\shared\etargets\etargets
C:\Program Files\MATLAB\R2006b\toolbox\shared\etargets\rtdxblks
C:\Program Files\MATLAB\R2006b\toolbox\rtw\targets\tic6000\tic6000
C:\Program Files\MATLAB\R2006b\toolbox\rtw\targets\tic6000\tic6000blks
C:\Program Files\MATLAB\R2006b\toolbox\rtw\targets\tic6000\
tic6000demos
C:\Program Files\MATLAB\R2006b\toolbox\vipblks\vipblks
C:\Program Files\MATLAB\R2006b\toolbox\vipblks\vipmasks
C:\Program Files\MATLAB\R2006b\toolbox\vipblks\vipmex
C:\Program Files\MATLAB\R2006b\toolbox\vipblks\vipdemos
C:\Program Files\MATLAB\R2006b\toolbox\vr\vr
C:\Program Files\MATLAB\R2006b\toolbox\vr\vrdemos
C:\Program Files\MATLAB\R2006b\toolbox\wavelet\wavelet
C:\Program Files\MATLAB\R2006b\toolbox\wavelet\wavedemo
C:\Program Files\MATLAB\R2006b\toolbox\rtw\targets\xpc\xpc
C:\Program Files\MATLAB\R2006b\toolbox\rtw\targets\xpc\target\build\
xpcblocks
C:\Program Files\MATLAB\R2006b\toolbox\rtw\targets\xpc\xpcdemos
C:\Program Files\MATLAB\R2006b\toolbox\rtw\targets\xpc\xpc\xpcmngr
C:\Program Files\MATLAB\R2006b\work
C:\Program Files\MATLAB\R2006b\toolbox\physmod\network_engine\
network_engine
C:\Program Files\MATLAB\R2006b\toolbox\physmod\network_engine\
ne_sli
C:\Program Files\MATLAB\R2006b\toolbox\physmod\network_engine\
library
C:\Program Files\MATLAB\R2006b\toolbox\physmod\sh\sh
C:\Program Files\MATLAB\R2006b\toolbox\physmod\sh\shdemos
C:\Program Files\MATLAB\R2006b\toolbox\physmod\sh\library

To find out what version of MATLAB, Simulink, and its toolboxes you are using, type the command

≫ **ver**

It will provide you with version information on MATLAB, Simulink, and all associated software as follows.

MATLAB Version 7.3.0.267 (R2006b)
MATLAB License Number: DEMO
Operating System: Microsoft Windows XP Version 5.1 (Build 2600: Service Pack 2)
Java VM Version: Java 1.5.0 with Sun Microsystems Inc. Java HotSpot(TM) Client VM mixed mode

----

MATLAB	Version 7.3	(R2006b)
Simulink	Version 6.5	(R2006b)
Aerospace Blockset	Version 2.2	(R2006b)
Aerospace Toolbox	Version 1.0	(R2006b)
Bioinformatics Toolbox	Version 2.4	(R2006b)
Communications Blockset	Version 3.4	(R2006b)
Communications Toolbox	Version 3.4	(R2006b)
Control System Toolbox	Version 7.1	(R2006b)
Curve Fitting Toolbox	Version 1.1.6	(R2006b)
Data Acquisition Toolbox	Version 2.9	(R2006b)
Database Toolbox	Version 3.2	(R2006b)
Datafeed Toolbox	Version 1.9	(R2006b)
Embedded Target for Infineon C166 Microcontrollers	Version 1.3	(R2006b)
Embedded Target for Motorola MPC555	Version 2.0.5	(R2006b)
Embedded Target for TI C2000 DSP(tm)	Version 2.1	(R2006b)
Embedded Target for TI C6000 DSP(tm)	Version 3.1	(R2006b)
Excel Link	Version 2.4	(R2006b)
Extended Symbolic Math Toolbox	Version 3.1.5	(R2006b)
Filter Design HDL Coder	Version 1.5	(R2006b)
Filter Design Toolbox	Version 4.0	(R2006b)
Financial Derivatives Toolbox	Version 4.1	(R2006b)
Financial Toolbox	Version 3.1	(R2006b)
Fixed-Income Toolbox	Version 1.2	(R2006b)
Fixed-Point Toolbox	Version 1.5	(R2006b)
Fuzzy Logic Toolbox	Version 2.2.4	(R2006b)
GARCH Toolbox	Version 2.3	(R2006b)
Gauges Blockset	Version 2.0.4	(R2006b)
Genetic Algorithm Direct Search Toolbox	Version 2.0.2	(R2006b)
Image Acquisition Toolbox	Version 2.0	(R2006b)
Image Processing Toolbox	Version 5.3	(R2006b)
Instrument Control Toolbox	Version 2.4.1	(R2006b)
Link for Code Composer Studio	Version 2.1	(R2006b)
Link for ModelSim	Version 2.1	(R2006b)
Linkfor TASKING	Version 1.0.1	(R2006b)

MATLAB Report Generator	Version 3.1	(R2006b)
Mapping Toolbox	Version 2.4	(R2006b)
Model Predictive Control Toolbox	Version 2.2.3	(R2006b)
Model-Based Calibration Toolbox	Version 3.1	(R2006b)
Neural Network Toolbox	Version 5.0.1	(R2006b)
OPC Toolbox	Version 2.0.3	(R2006b)
Optimization Toolbox	Version 3.1	(R2006b)
Partial Differential Equation Toolbox	Version 1.0.9	(R2006b)
RF Blockset	Version 1.3.1	(R2006b)
RF Toolbox	Version 2.0	(R2006b)
Real-Time Windows Target	Version 2.6.2	(R2006b)
Real-Time Workshop	Version 6.5	(R2006b)
Real-Time Workshop Embedded Coder	Version 4.5	(R2006b)
Robust Control Toolbox	Version 3.1.1	(R2006b)
Signal Processing Blockset	Version 6.4	(R2006b)
Signal Processing Toolbox	Version 6.6	(R2006b)
SimBiology	Version 2.0.1	(R2006b)
SimDriveline	Version 1.2.1	(R2006b)
SimEvents	Version 1.2	(R2006b)
SimHydraulics	Version 1.1	(R2006b)
SimMechanics	Version 2.5	(R2006b)
SimPowerSystems	Version 4.3	(R2006b)
Simulink Accelerator	Version 6.5	(R2006b)
Simulink Control Design	Version 2.0.1	(R2006b)
Simulink Fixed Point	Version 5.3	(R2006b)
Simulink HDL Coder	Version 1.0	(R2006b)
Simulink Parameter Estimation	Version 1.1.4	(R2006b)
Simulink Report Generator	Version 3.1	(R2006b)
Simulink Response Optimization	Version 3.1	(R2006b)
Simulink Verification and Validation	Version 2.0	(R2006b)
Spline Toolbox	Version 3.3.1	(R2006b)
Stateflow	Version 6.5	(R2006b)
Stateflow Coder	Version 6.5	(R2006b)
Statistics Toolbox	Version 5.3	(R2006b)
Symbolic Math Toolbox	Version 3.1.5	(R2006b)
System Identification Toolbox	Version 6.2	(R2006b)
SystemTest	Version 1.0.1	(R2006b)
Video and Image Processing Blockset	Version 2.2	(R2006b)
Virtual Reality Toolbox	Version 4.4	(R2006b)
Wavelet Toolbox	Version 3.1	(R2006b)
xPC Target	Version 3.1	(R2006b)

Your Simulink preferences are set using the MATLAB **Preferences** dialog box (Fig. 8.4). To open this dialog box, select **Preferences** from the Simulink **File** menu.

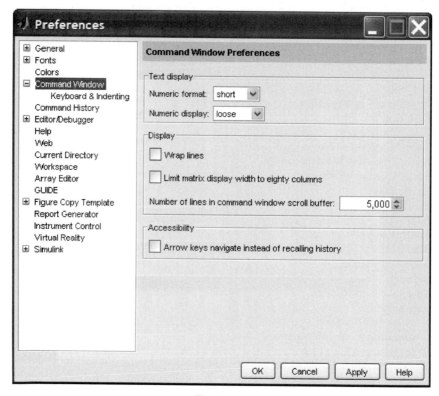

Fig. 8.4

A variety of Simulink demonstrations can be accessed using the **demo** command. Simulink demonstrations can always be started by selecting **Demos** from the **Help** pull-down menu. Try typing the command

≫**demo**

This results in the interface window shown in Fig. 8.5. Note that the Simulink demonstrations are contained under **+ Simulink**.

Fig. 8.5

8.5 Conclusion

A variety of other Simulink block libraries and tools are likely available on your system. Take a look, and start exploring! In the next lecture we will start constructing Simulink models.

Practice Exercises

8.1   Load an existing Simulink model and do some simple analysis using it. This example is a model of the one-dimensional (vertical) behavior of an F-14 fighter.

Type the following command into the **MATLAB Command Window** to load the data required by the model:

>> **f14dat;**

Next load the system and associated numerical constants:

>> **f14;**

The loaded model should look like the one in Fig. 8.6.

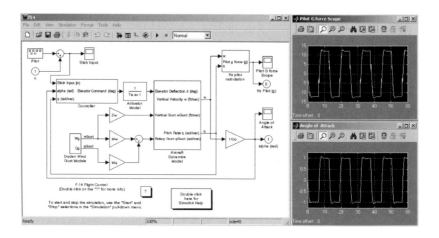

Fig. 8.6

Double-click on the **Stick Input Scope** so that inputs and outputs all are visible. Go to **Simulation** and use the pull-down window to select **Start**. See what happens. Try using the **Autoscale** (binoculars) and **Zoom** functions (magnifying glass) in the **Scope** windows. Try other **Pilot** inputs by opening the **Pilot Signal Generator**. Try **sine**, **sawtooth**, and **random** inputs. Try dragging other blocks from the **Simulink Library Browser** onto the diagram. For example, replace the **Pilot Signal Generator** with other **Sources** and execute the simulation again.

As you initially open the model, the input excitation should be as shown in Fig. 8.7.

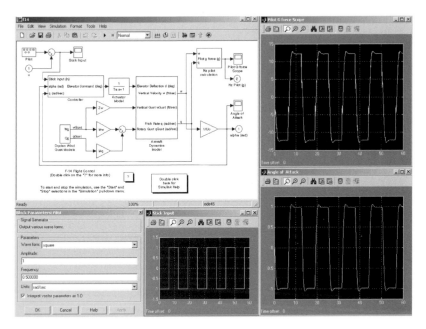

Fig. 8.7

Notes

Building a Simple Simulink® Model

9.1 Introduction and Objectives

This chapter provides the reader with an example of how to build a Simulink® model. Further examples are given in the following chapters. This chapter will provide the reader with the tools to start building simple Simulink models.

Upon completion of this chapter, the reader will be able to generate simple Simulink models, execute and simulate simple systems implemented in Simulink, and modify Simulink model block parameters.

9.2 Population Model

Our first Simulink system models a population in which the number of members over time follows the equation

$$dm/dt = a^*m - b^*m^*m$$

In the previous equation, the variable a is taken to represent the reproductive rate and b represents competition. We will construct the model shown in Fig. 9.1.

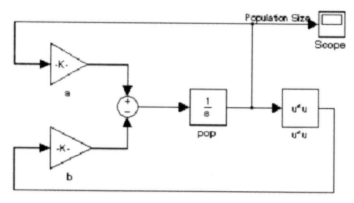

Fig. 9.1

We will name this Simulink model **pops.mdl**. Then 20 simulations will be run using different starting populations.

To create the model, first enter **Simulink** in the MATLAB® **Command Window**. On Microsoft Windows, the **Simulink Library Browser** appears as shown in Fig. 9.2.

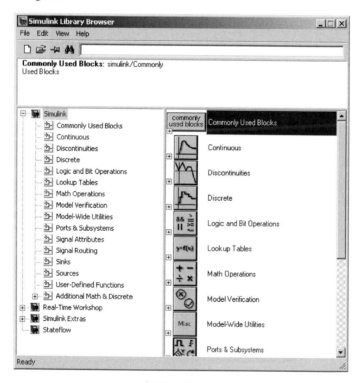

Fig. 9.2

To create a new model in Windows, click the **New Model** button on the Library Browser's toolbar (see Fig. 9.3).

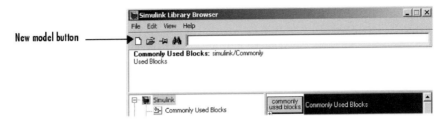

New model button

Fig. 9.3

Simulink opens a new model window (Fig. 9.4).

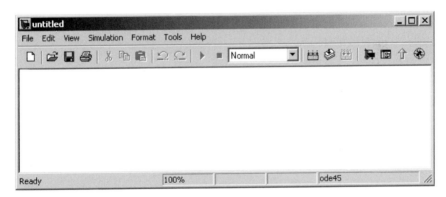

Fig. 9.4

Use **File, Save as . . .** to rename the file to **pops.mdl**. Note that for Simulink to recognize this as a Simulink model the filename must end with a **.mdl**. The full name of your file should then be **pops.mdl**. The name in the upper left corner of your Simulink model window will now say **pops**.

You can next start to populate your Simulink model with the required block elements. The model elements are two gains, an integrator, and a squaring element. A scope will be used to see the results during each run. All elements will finally be connected together (see Figs 9.5–9.7).

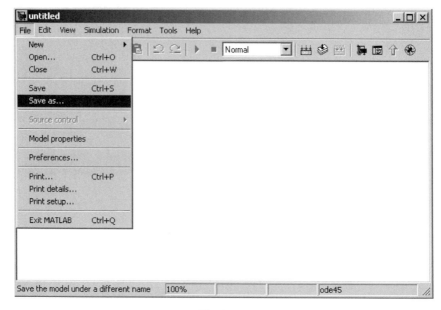

Fig. 9.5

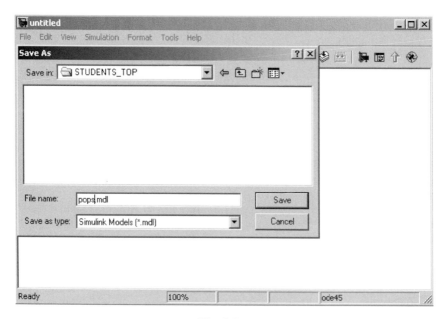

Fig. 9.6

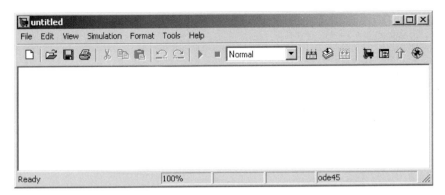

Fig. 9.7

We are now ready to start collecting our model elements and placing them into the **pops** window. Go first to the **Simulink Library Browser** and select the **Continuous** library. Next select the **Integrator** block, and while holding down the left mouse button, drag the block over into your **pops** window. The operation should appear as in the following figures.

Note that an asterisk * appears after **pops** in the upper left corner of the window after an unsaved modification is made. This tells you that the Simulink model has been changed since you last saved the file (see Figs 9.8–9.10).

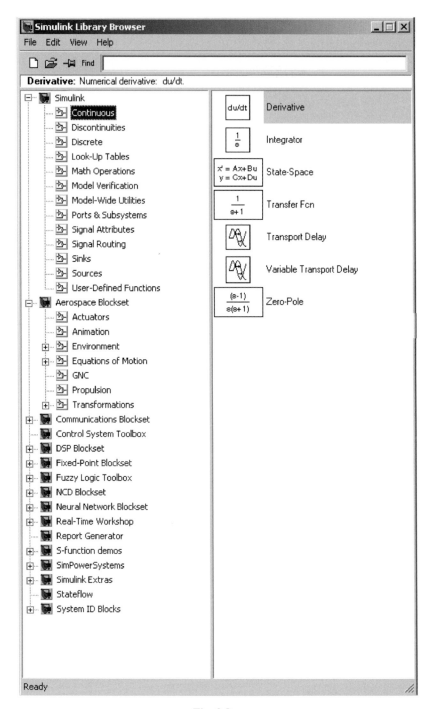

Fig. 9.8

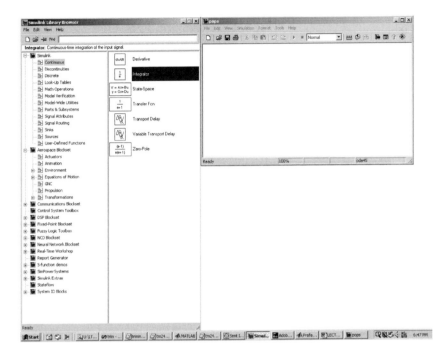

Fig. 9.9

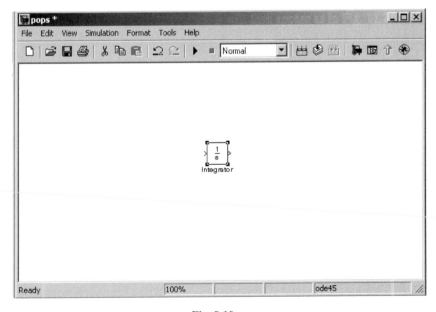

Fig. 9.10

You next need to add a couple of gains. Open the **Math Operations** library as shown in Fig. 9.11.

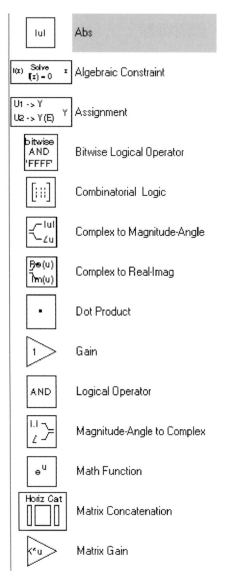

Fig. 9.11

Select the **Gain** block and drag it into the **pops** window. Repeat this operation a second time to generate the second gain element. The **pops** window will appear as shown in Figs 9.12 and 9.13.

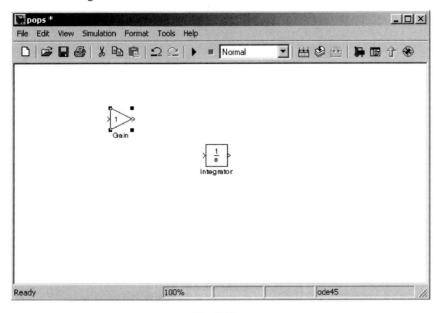

Fig. 9.12

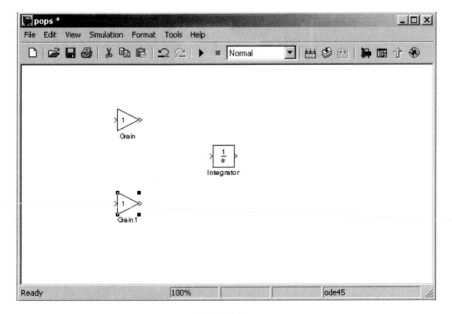

Fig. 9.13

Now you need a summing junction. You will find this near the bottom of the
Math Operations library (see Figs 9.14 and 9.15).

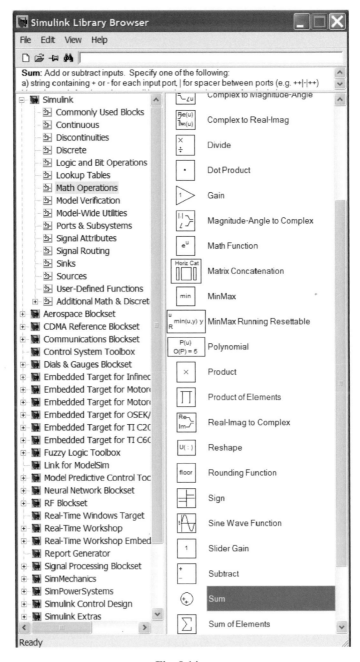

Fig. 9.14

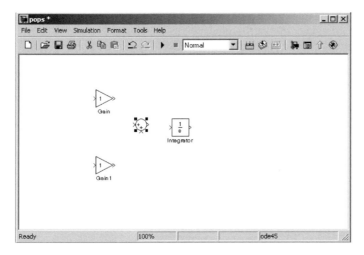

Fig. 9.15

Now we need to add the element to square the signal. We select the **Math Function** block from the library of **Math Operations**. We again drag this element into our **pops** model. You might consider saving your model again every now and then during this model-generation process. You would hate to have to start again from scratch if something bad happened during the construction process (see Fig. 9.16).

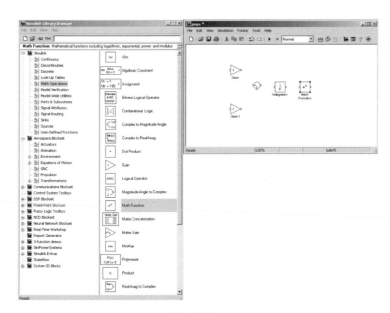

Fig. 9.16

To view the output of this dynamic system, a **Scope** output block is required. This is available from the Simulink library of **Sinks**.

Again, select the **Scope** using the mouse. Drag the **Scope** until it is located at the desired spot in the pops model window. Release the **Scope** at this location (see Fig. 9.17).

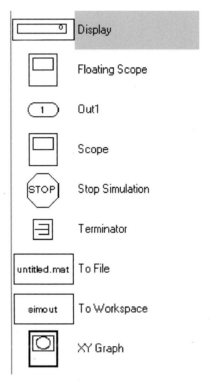

Fig. 9.17

Our model now (we hope) looks like the window shown in Fig. 9.18.

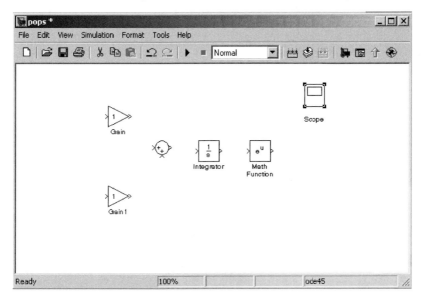

Fig. 9.18

We next need to connect all the elements that we generated. This is done by mouse selection of the output ports of our block elements and connecting them to the input ports of the other block elements (see Fig. 9.19).

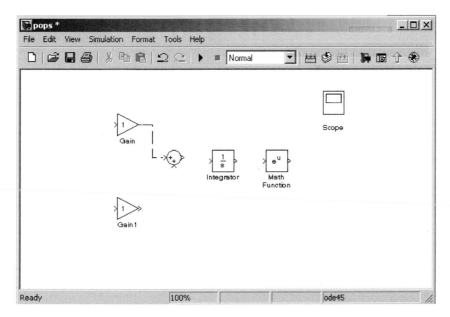

Fig. 9.19

A helpful hint might appear on the screen as follows. Simulink attempts to aid you through the connection process.

Remember to use left mouse clicks to select the starting ports, and make sure to position the mouse close enough to the final in port for Simulink to recognize the connection. You may need a little practice to get used to making these connections. The dashed lines show connections in progress. Final connections are shown by solid arrows (see Figs 9.20–9.22).

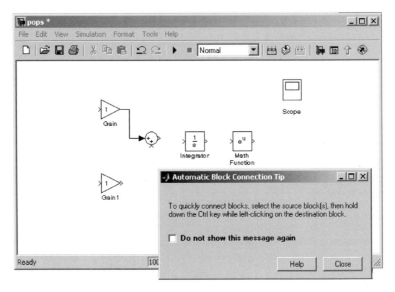

Fig. 9.20

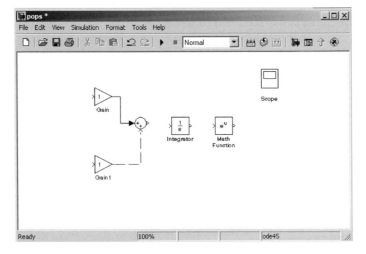

Fig. 9.21

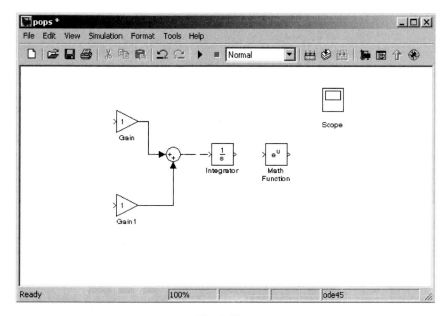

Fig. 9.22

Connections are made one at a time (see Figs 9.23 and 9.24).

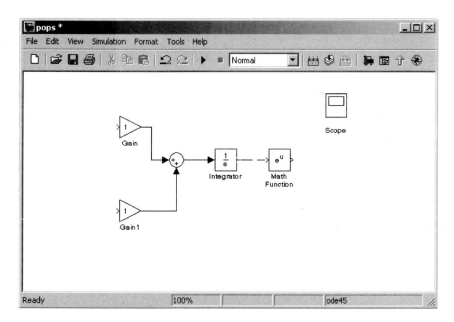

Fig. 9.23

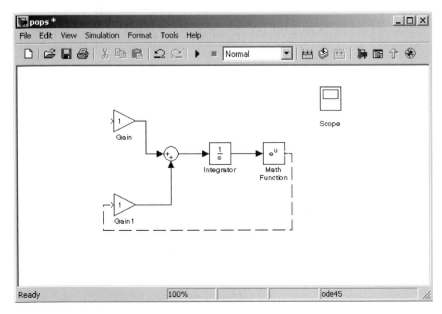

Fig. 9.24

Hold the right mouse button to branch from a connecting line (see Figs 9.25 and 9.26).

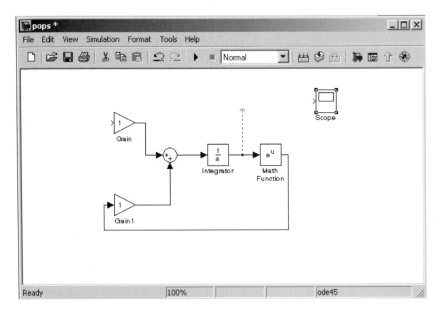

Fig. 9.25

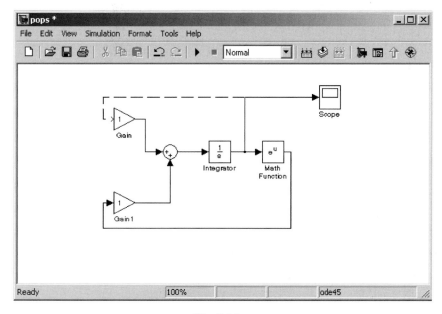

Fig. 9.26

We are now branching from the corner of an existing line (see Fig. 9.27).

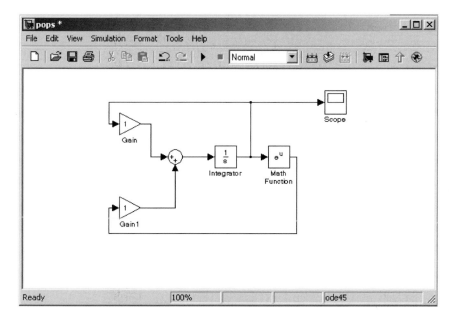

Fig. 9.27

By double-clicking on an element we wish to modify, we can open the appropriate parameter window. Here we will change the sign of the second summation into a subtraction. In the list of signs, the second **+** is deleted and a **−** sign is added. The next window shows the changed parameter. When **Apply** is selected, the sign is changed in the **pops** model diagram. Note that the | element in the **List of Signs** denotes a spacer that later will be moved to change the appearance of the **Sum** (see Figs 9.28–9.30).

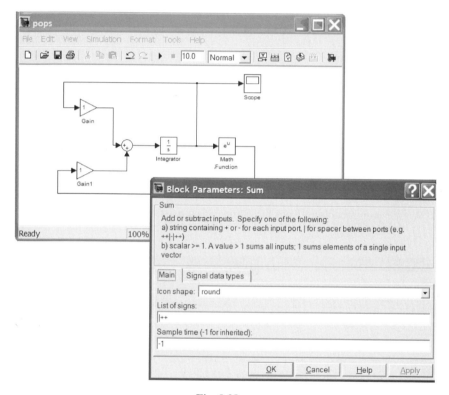

Fig. 9.28

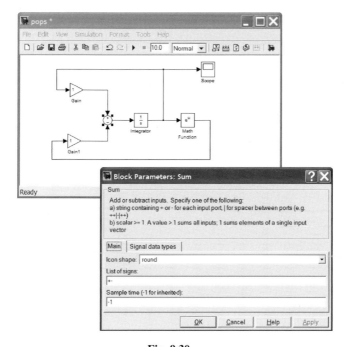

Fig. 9.29

Fig. 9.30

We next double-click on the **Math Function** block to change it into a squaring block (see Fig. 9.31).

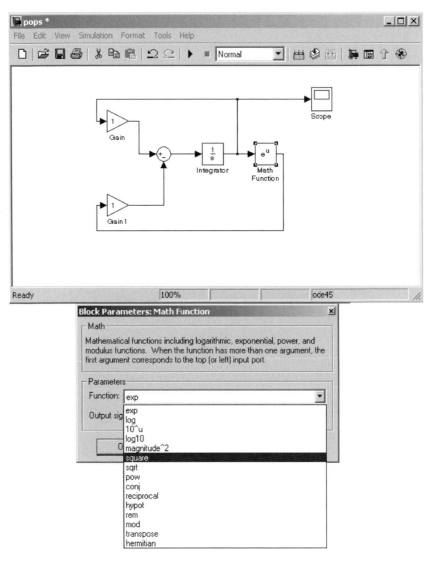

Fig. 9.31

We then type over the name **Math Function** to give it the new name **u*u** (see Fig. 9.32).

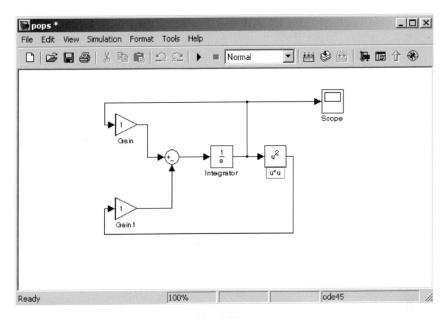

Fig. 9.32

The final changes to the elements are to give the first gain we created the block title **a** and assign it a value of **0.02309**. Then we change the second block title to **b** and give it a value of **0.0005157**. Again, double-click on the elements to change the parameter values. You select the gain element titles to rename them.

These operations are shown in Fig. 9.33.

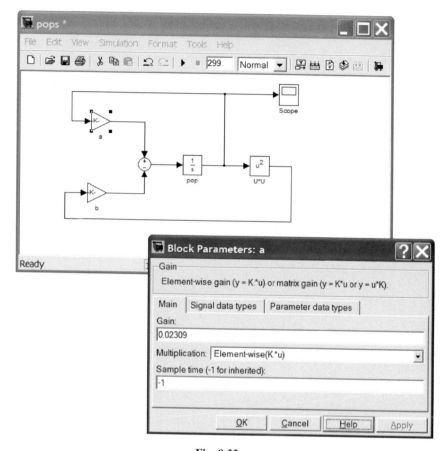

Fig. 9.33

Here we typed in the first gain parameter value of **0.02309**.
This value is entered into the gain block by selecting **Apply** (see Fig. 9.34).

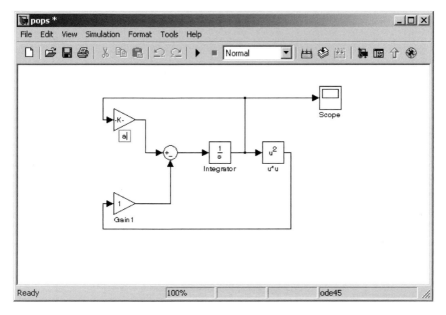

Fig. 9.34

Next the name **Gain** must be selected and changed to the block name **a**.

To change the name of **Gain**, use the mouse to select the text, and then type over the old name with the new name.

The previous process is repeated after selecting the **Gain1** element. This process is summarized in Fig. 9.35.

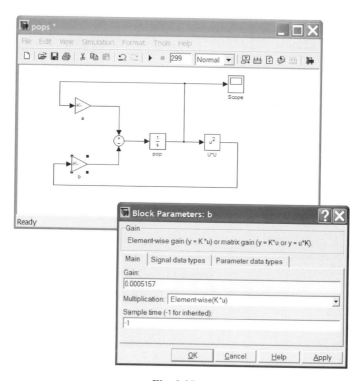

Fig. 9.35

Here the value of the second gain element is entered. After this value is applied, the block label is changed into **b**. The result of these actions is the modified model in Fig. 9.36.

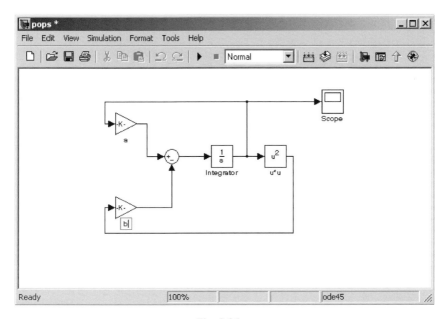

Fig. 9.36

We can now open up our **Scope** and run our population simulation (see Fig. 9.37).

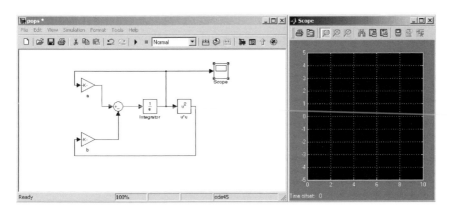

Fig. 9.37

To more closely match our block diagram, we will make an additional change to our **Sum**. The entry into the summation junction is now modified to match our original diagram by removing the spacer at the start of the list of signs. This could also have been done earlier when we made our change in sign (see Figs 9.38 and 9.39).

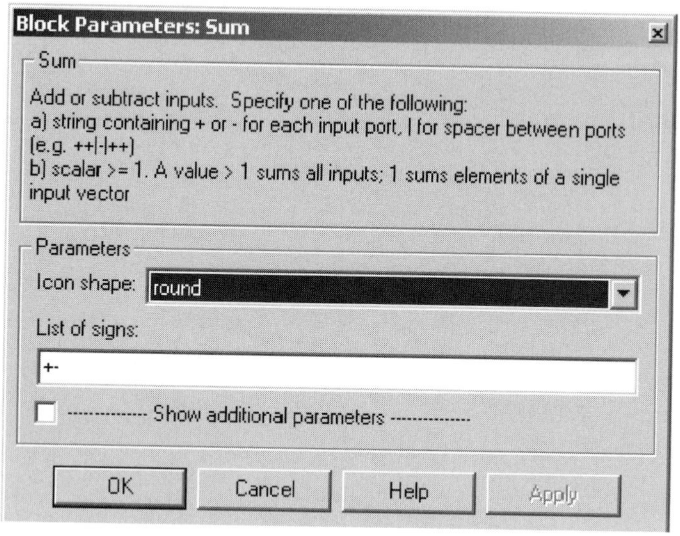

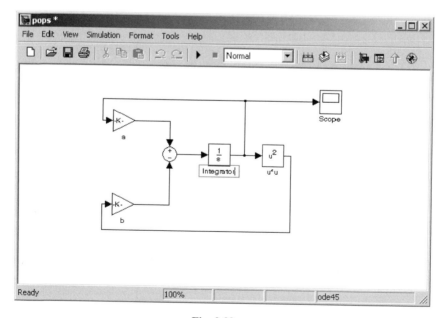

Fig. 9.38

Fig. 9.39

To prepare the model for use in Sec. 9.3 we will select the name **Integrator** and change it to **pop**. We will give it an initial condition of 20.0. The initial condition is stored under **Model Properties** (see Fig. 9.40).

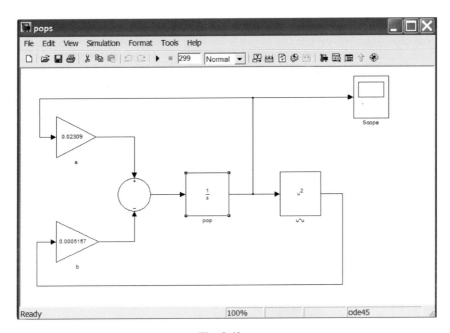

Fig. 9.40

Double-clicking on the **pop** integrator with the mouse opens the **Block Parameters** menu shown in Fig. 9.41.

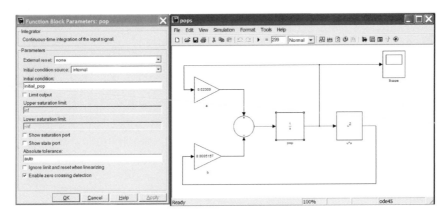

Fig. 9.41

Note that **Model Properties** is selected from the **pops** window using the right mouse button menu. Type **initial_pop = 20.0;** into the **Callbacks, Model pre-load function** window. Note that preloaded data are only placed into memory when the model is opened. The advantage of this is that the values can be updated at any time from MATLAB or from within the Simulink model without being overwritten by the mode defaults. The disadvantage is that while the model is being created these data are not entered into memory. You must both close and then reopen the model to preload the data, or you must manually enter it into the MATLAB **Command Window** during the model-creation process (see Fig. 9.42).

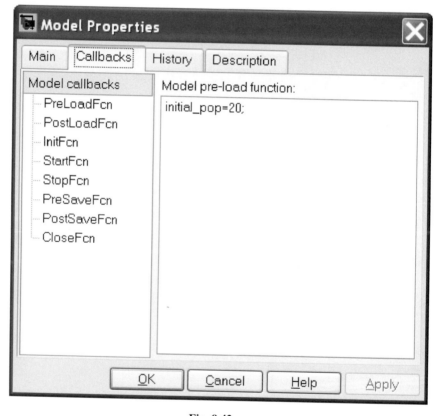

Fig. 9.42

This value can also be set and reset by typing the same line into the MATLAB **Command Window**. We will update this value in Sec. 9.3 using this feature.

From the **Simulation** pull-down in the **pops** window, select **Configuration Parameters**. This can also be accomplished using the **Ctrl-E** shortcut. Modify the **Solver** to appear as in Fig. 9.43.

Similarly, modify the **Workspace I/O** to appear as in Fig. 9.43. Check the radio button in front of the fields that need to be modified but are not currently active. After making the modifications, deselect the field.

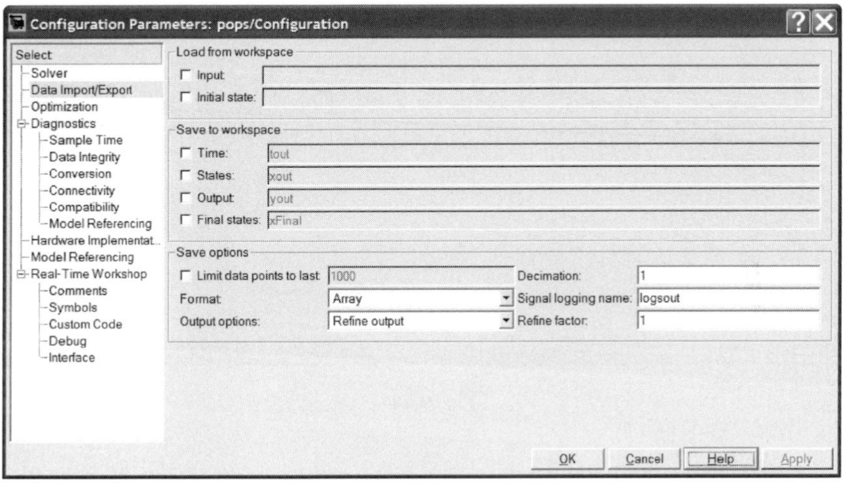

Fig. 9.43

The **Scope** plot in Fig. 9.44 shows the response of the system after **Simulation, Start** is selected from the pull-down menu. As a shortcut, you can use the triangle symbol to start the simulation.

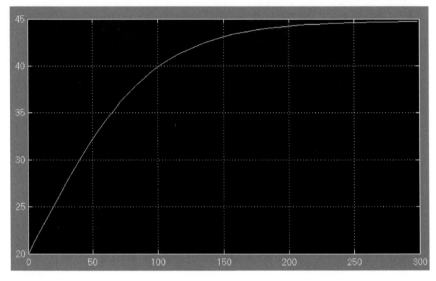

Fig. 9.44

Note that if you do not get the desired plot scaling, you can use the binocular symbol to autoscale the plot to the data. You can also use the magnifying glasses to zoom into and out of the figure.

9.3   Analyzing the Population Model

In this section we will make multiple runs with different initial populations using the model developed in Sec. 9.2. We have completed our system block diagram and are now ready to run it. For a first look, double-click on the **Scope** to open it. Then select **Simulation, Start** to execute it.

The following M-file runs our **pops** Simulink model through 20 simulations and plots the results. Use the Runge-Kutta integration method to simulate **pops** until t = 300.

```
%Population Simulation M-file
clf
for ind = 20:-1:1
   initial_pop = ind*5 + 1;
   [t,x] = sim('pops',300);
   plot(t,x);
   hold on;
end
```

```
%
axis([0,300,0,100]);
xlabel('Time');
ylabel('Population Size');
```

Different numbers in **a** and **b** will result in different initial and final populations. You can change these values and run the simulation again. Do this by simply opening the gain elements in the model (see Fig. 9.45).

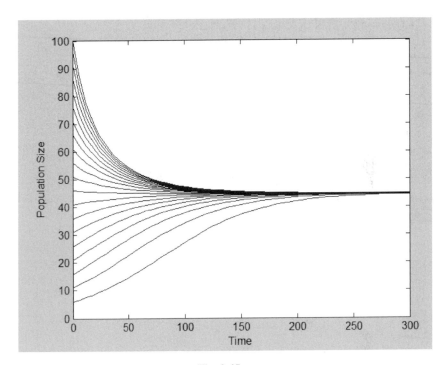

Fig. 9.45

9.4 Conclusion

This chapter provides a first example of how to build a Simulink model. Further examples are given in the following chapters.

You are now ready to start building simple Simulink models on your own! Try modifying the **pops.mdl** file you just created. Take the input line to the scope and give it a name, like **Population Size**. This is done by double-clicking on the signal line. Rearrange the lines to look more like the original diagram. Rearrange the format of the summation to look more like the original. Hook up scopes to other parts of the diagram and look at the results.

Practice Exercises

9.1   The exercise problem is to generate a Simulink model of a forced pendulum. The model of the pendulum is given in Fig. 9.46.

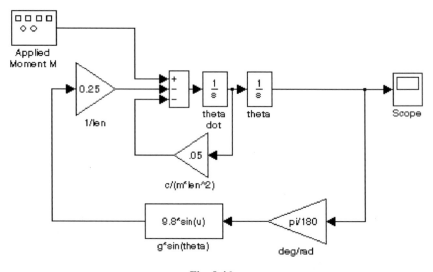

Fig. 9.46

The forcing function for this model follows. It is a **square** wave with a frequency of 0.1 radians/second and amplitude of 1. Initialize **theta dot** to 0 degrees/second and **theta** to 45 deg. The resulting output is periodic with the waveform show on the scope in Fig. 9.47.

After plotting the square wave response, change the excitation function to a **Pulse Generator**. Set it to get the same response (use the same amplitude and period). Next try out other excitations and **Sources**.

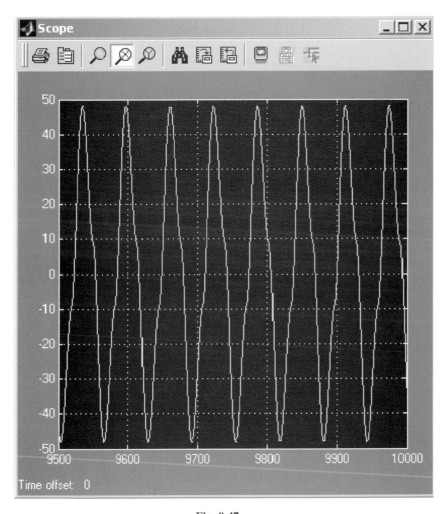

Fig. 9.47

Notes

10
Building Simulink® Linear Models

10.1 Introduction and Objectives

This chapter provides a brief introduction to using Simulink® and its library of models and other elements for dynamic system modeling.

Upon completion of this chapter, the reader will be able to build simple dynamic single-input, single-output Simulink models using transfer functions and execute and simulate simple systems implemented in Simulink.

10.2 Transfer Function Modeling in Simulink®

This system demonstrates the use of a transfer function to model a dynamic system. The transfer function we will examine is

$$\frac{2s + 1}{2s^2 + 5s + 3}$$

We will call our model **TF.mdl**. Open the Simulink **Library Browser** using the MATLAB® command **simulink** from the MATLAB **Command Window**. Then, from the Simulink **Library Browser**, select **File**, **New**, and then **Model**. Note from the following figure that **Ctrl+N** can be used as a shortcut to accomplish this. A new, untitled model window is then opened. Save this model under the name **TF.mdl** using **File**, **Save as** . . . as shown in Figs 10.1–10.3.

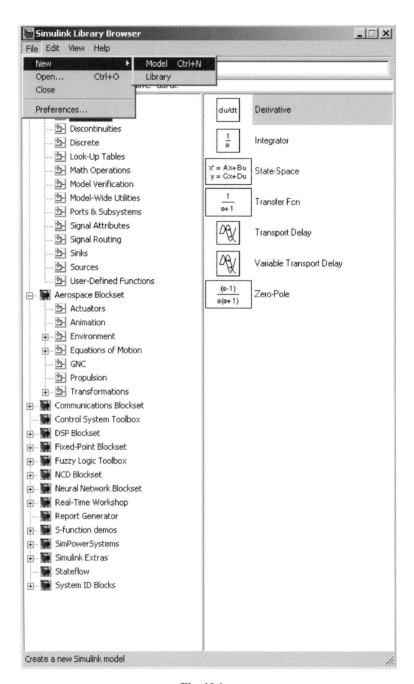

Fig. 10.1

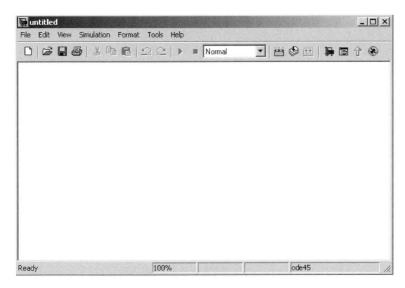

Fig. 10.2

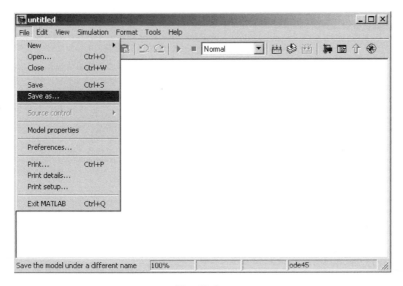

Fig. 10.3

We will want to add three elements to our model window. The model itself will require a **Transfer Function** block. This model will require an input and an output. The input will be a **Function Generator**. The output will be a **Scope**.

Transfer Fcn block:

Signal Generator block:

Scope block:

The transfer function properties will be as shown in Fig. 10.4.

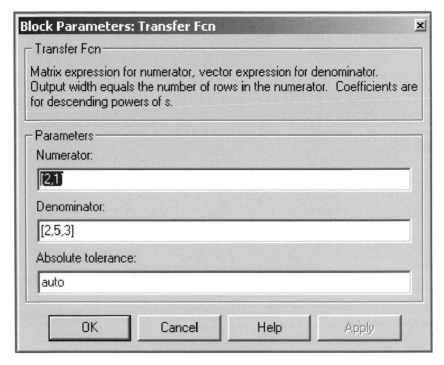

Fig. 10.4

The output of the **Function Generator** will be as shown in Fig. 10.5.

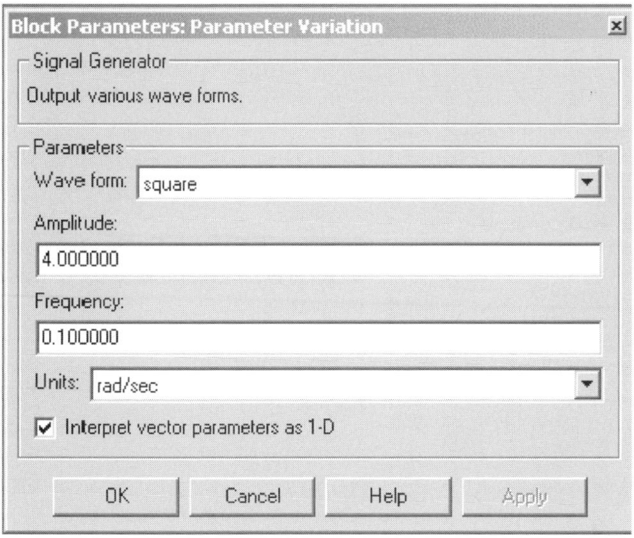

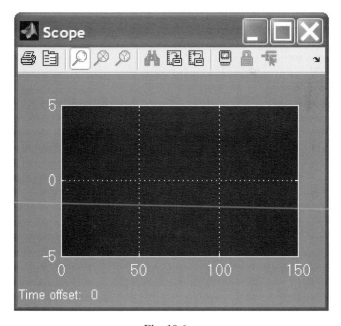

Fig. 10.5

The scope will simply show the system output for the requested run time. Figure 10.6 is a **Scope** with the **Simulation Parameters** set at 150 seconds.

Fig. 10.6

We will get the **Transfer Function** block from the library of **Continuous** systems under the name **Transfer Fcn**. The list of elements within this library is shown in Fig. 10.7.

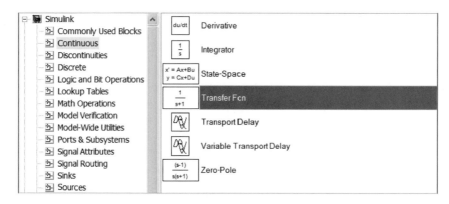

Fig. 10.7

We will get the input **Signal Generator** from the library of **Sources**. This list is given in Fig. 10.8.

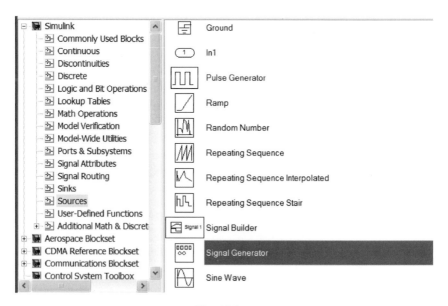

Fig. 10.8

Finally, the output **Scope** will be taken from the library of **Sinks**. This library is shown in Fig. 10.9.

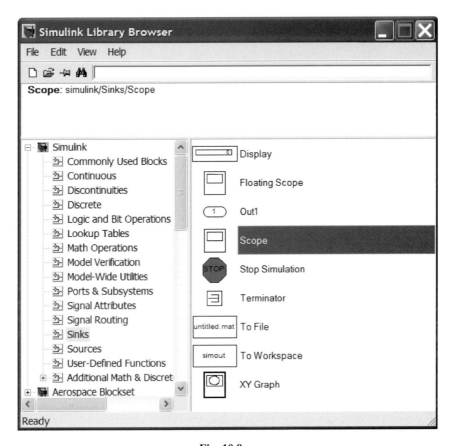

Fig. 10.9

This system is next constructed by dragging these three blocks from their library into the **TF** Simulink model window. After you perform this action, the window should appear as in Fig. 10.10.

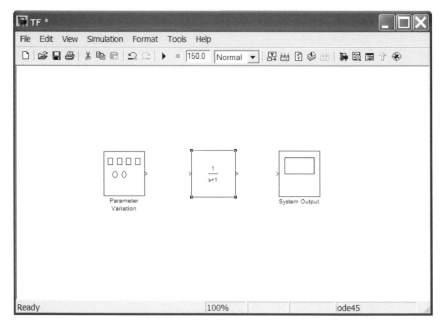

Fig. 10.10

Next connect the **Signal Generator** block to the **Transfer Fcn** block. Repeat this action by either using the mouse to drag the output arrow from the **Transfer Fcn** block to the **Scope** or by clicking on the **Transfer Fcn** block while holding the **Ctrl** key and then selecting the **Scope** as a shortcut. After you do this action, your model should appear as in Fig. 10.11. This might be a good time to save your work (note the * following the name **TF** in the upper left corner, which denotes changes were made to the model since it was last saved).

Note that to generate the following two views I first used the mouse drag and connect method. Then, for my second connection, I used the **Ctrl** key shortcut and let Simulink draw the line for me. When constructing complex block diagrams, you can modify the connections by selecting and moving them just like any library element.

Next let us rename the three elements. Click on the first title (**Signal Generator**) and change the name to **Parameter Variation**. Then click on the second title (**Transfer Fcn**) and change the name to **System Under Test**. Finally, click on the third title (**Scope**) and change the name to **System Output**. Note that I have right-clicked on the name **System Under Test** and dragged it to its new location above the block. The result will look as in Fig. 10.11.

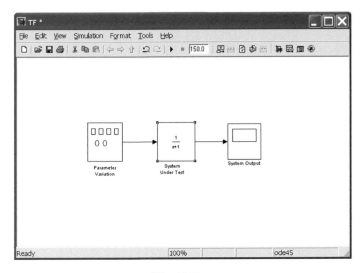

Fig. 10.11

We next need to modify the **Parameter Variation** input. Double-click on the block and input the values shown in Fig. 10.12. Then **Apply** these values and close the window. The diagram will not appear to have changed.

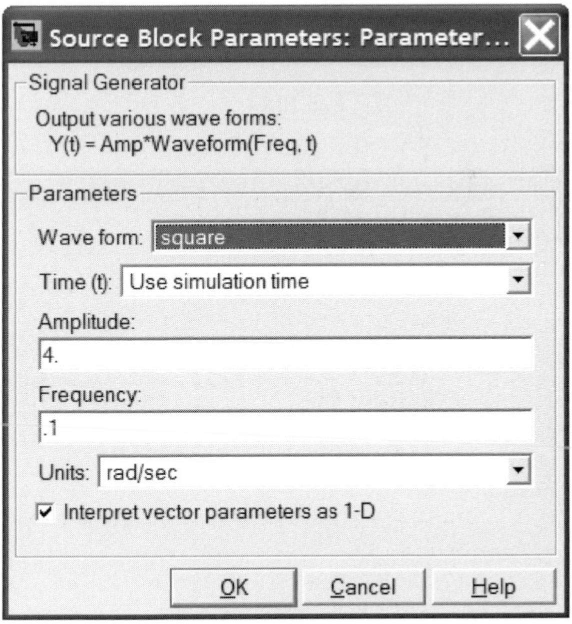

Fig. 10.12

Next input the desired transfer function. Double-click on the block and input the values shown in Fig. 10.13. Then **Apply** these values and close the window. The diagram will change to reflect the new transfer function model. This new model is shown in Fig. 10.14. Note that either spaces or commas could be used to separate the constant coefficients in the numerator and in the denominator. This is exactly the same as when using the MATLAB **Command Window** or when programming an M-file.

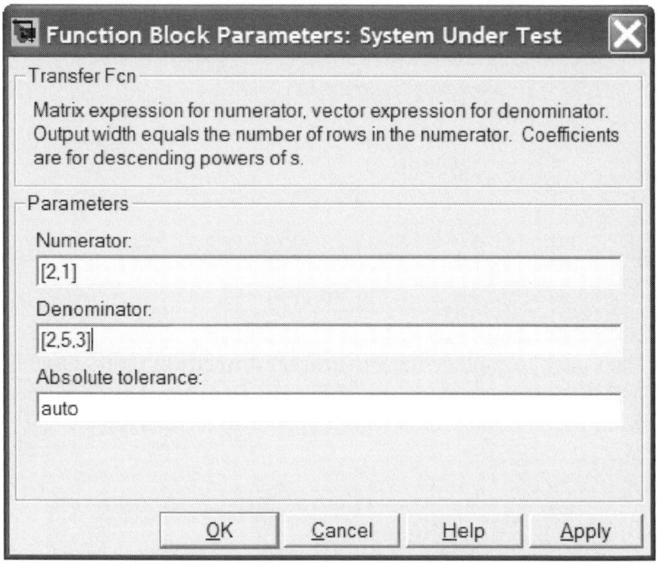

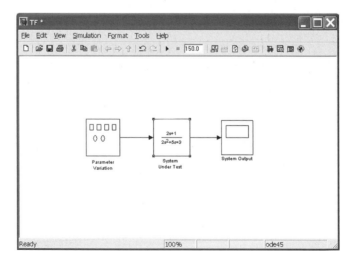

Fig. 10.13

Fig. 10.14

We are now ready to analyze this model. Select **Simulation**, then **Start**. The square wave will be input into the **System Under Test** (see Fig. 10.15).

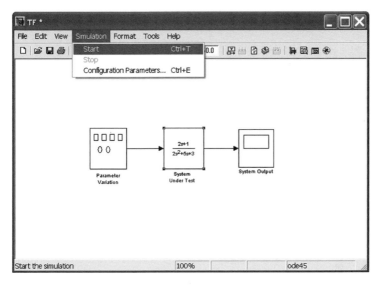

Fig. 10.15

Next open the **Scope**. You will see the time history of this response. You might need to use the autoscale option (binoculars symbol) to better view this response. The response should be as shown in Fig. 10.16.

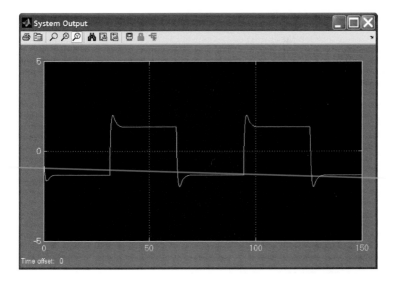

Fig. 10.16

Try using the variety of excitation methods available in the **Simulink Signal Generator**, and examine the results using the **Scope**. Then compare these responses with those generated by the other **Source** blocks available within Simulink.

10.3   Zero-Pole Model

In this section we will generate the equivalent **Zero-Pole** model to the transfer function model we have been studying:

$$\frac{2s + 1}{2s^2 + 5s + 3}$$

This is accomplished using the MATLAB command **tf2zp**.

First enter the numerator and denominator coefficients for our model into the MATLAB **Command Window**:

> $\gg$ **num = [2,1];**
> $\gg$ **den = [2,5,3];**

Next we will do a transfer function to zero-pole conversion.

The command **[Z, P, K]=tf2zp(num,den)** finds the zeros, poles, and gains

$$H(s) = \frac{K(s - z1)(s - z2)(s - z3)}{(s - p1)(s - p2)(s - p3)}$$

from a transfer function in polynomial form:

$$H(s) = \frac{NUM(s)}{DEN(s)}$$

The vector **DEN** specifies the coefficients of the denominator in descending powers of s, and **NUM** indicates the numerator coefficients with as many rows as there are outputs. The zero locations are returned within the columns of the matrix **Z**; **Z** has as many columns as there are rows in **NUM**. The pole locations are returned within the column vector **P**, and the gains for each numerator transfer function are stored within the vector **K**.

Enter the following into the MATLAB **Command Window**:

> $\gg$ **[Z, P, K]=tf2zp(num,den)**

MATLAB responds with

Z =

   −0.5000

P =

   −1.5000

   −1.0000

K =

   1

The equivalent **Zero-Pole** model is

$$\frac{s + 0.5}{(s + 1.5)(s + 1.0)}$$

We will next modify our **TF.mdl** Simulink linear model. Save a copy of this model under the name **TFZPG.mdl**. Next delete the existing **Transfer Fcn** block. Obtain your **Zero-Pole** element from the Simulink **Library Browser**. This element is in the **Continuous Library**. Replace the **Transfer Fcn** block in the model **TF.mdl** with the **Zero-Pole** model modified with the model data generated in MATLAB. This results in the model shown in Fig. 10.17, saved as **TFZPG.mdl**. Note that this model gives you exactly the same results as are seen in Sec. 10.2 (see Fig. 10.18).

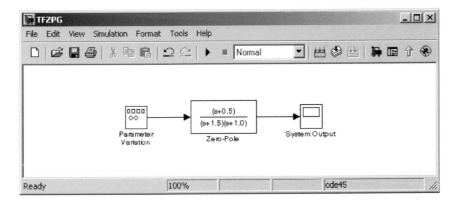

Fig. 10.17

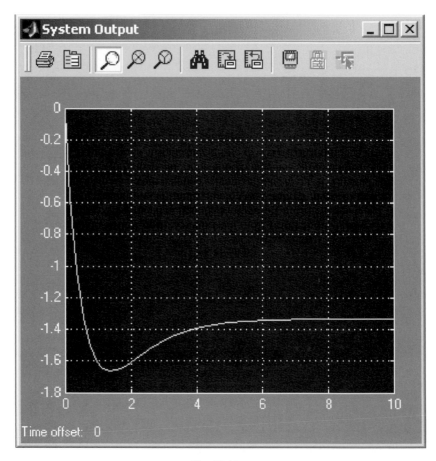

Fig. 10.18

10.4 State-Space Model

In this section we will generate the equivalent **State-Space** model to the transfer function model we have been studying:

$$\frac{2s + 1}{2s^2 + 5s + 3}$$

This is accomplished using the MATLAB command **tf2ss**, which does the transfer function to state-space conversion.

The MATLAB command **[A,B,C,D] = TF2SS(NUM,DEN)** calculates the state-space representation

$$\dot{x} = Ax + Bu$$

$$y = Cx + Du$$

of the system

$$H(s) = \frac{NUM(s)}{DEN(s)}$$

from a single input. The vector **DEN** must contain the coefficients of the denominator in descending powers of s. The matrix **NUM** must contain the numerator coefficients with as many rows as there are outputs y. The **A**, **B**, **C**, **D** matrices are returned in controller canonical form.

Enter into the MATLAB **Command Window** the following command:

$\gg$ **[A, B, C, D] = tf2ss(num, den)**

MATLAB returns with the equivalent **State-Space** model:

$$\dot{x} = Ax + Bu$$
$$y = Cx + Du$$

where for our example

A =

$\quad$ −2.5000 $\quad$ −1.5000

$\quad\quad$ 1.0000 $\quad\quad\quad$ 0

B =

$\quad$ 1

$\quad$ 0

C =

$\quad$ 1.0000 $\quad$ 0.5000

D =

$\quad$ 0

Replacing the **Transfer Fcn** block in the model **TF.mdl** with the **State-Space** model results in the model shown in Fig. 10.19.

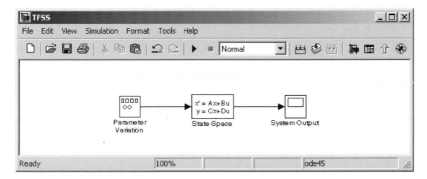

Fig. 10.19

Note that this model gives exactly the same results as are seen in Secs. 10.2 and 10.3 (see Fig. 10.20).

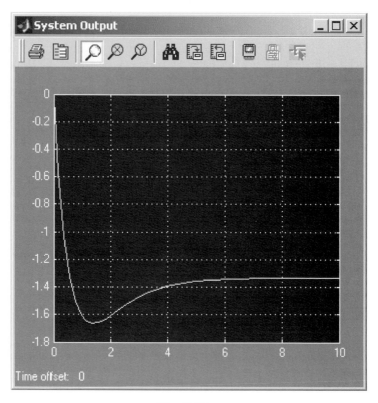

Fig. 10.20

10.5 Conclusion

This chapter was a very brief introduction to using Simulink and its library of **Transfer Fcn**, **Zero-Pole**, and **State-Space** models and other elements for dynamic system modeling. Further examples are given in the Simulink manual. In the following chapters and exercises, we will look at a variety of methods to design and analyze such systems in Simulink. We will also look at using these tools to build complex multiple-input, multiple-output (MIMO) systems.

Practice Exercises

10.1   This problem is to implement a transfer function in zero-pole form to model a dynamic system. The transfer function we will examine is

$$\frac{s^2 + 4s + 3}{2s^2 + 5s + 2}$$

Use the MATLAB command **tf2zp** to convert this into zero-pole form. Then implement this in Simulink. Simulate this model as in Chapter 10. Finally, convert this into **State-Space** form. Use the MATLAB command **tf2ss** to do this. Replace the element in your model with the **State-Space** element. Verify that the time response is the same.

The zero-pole diagram should resemble the window shown in Fig. 10.21.

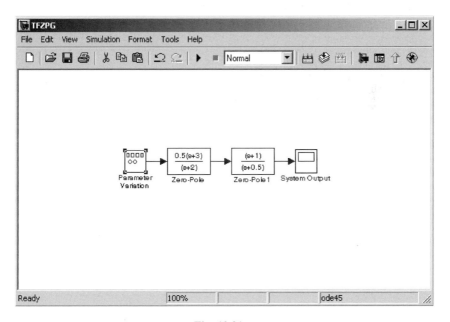

Fig. 10.21

Notes

11
LTI Viewer and SISO Design Tool

11.1 Introduction and Objectives

This chapter will introduce the reader to using the Simulink® **LTI Viewer** for the analysis of linear, time-invariant dynamic systems. It will also provide a very brief introduction to using the **SISO Design Tool** for the design of LTI systems.

Upon completion of this chapter, the reader will be able to design and analyze simple dynamic single-input, single-output (SISO) Simulink models using the **LTI Viewer** and the **SISO Design Tool**; become familiar with design and analysis methods, including step and other time domain responses, Bode diagrams, Nichols charts, Nyquist charts, and root analysis; and see how to use these tools for more complex multiple-input, multiple-output (MIMO) systems.

11.2 Introduction to the Simulink® LTI Viewer

To understand the use of the **LTI Viewer**, we will use the same transfer function model we examined in Chapter 10. This system will again be used to demonstrate the use of a transfer function to model a dynamic system. The transfer function we will study using the **LTI Viewer** is

$$\frac{2s + 1}{2s^2 + 5s + 3}$$

The **LTI Viewer** is a Simulink graphical user interface (GUI) for the analysis of linear, time-invariant (LTI) systems. The **LTI Viewer** is used to view and compare the response plots of SISO systems. It can also be used to examine SISO combinations in MIMO systems. Several linear models can be examined at the same time. You can generate time and frequency response plots as well as system roots to examine key response parameters. These include the system's rise time, maximum overshoot, and stability margins.

The **LTI Viewer** can display up to eight different plot types simultaneously: 1) **step response**, 2) **impulse response**, 3) **Bode diagrams** (either magnitude and phase or just magnitude), 4) **Nyquist charts**, 5) **Nichols charts**, 6) **sigma plot**, 7) **pole/zero plots**, and 8) **I/O pole/zero plots**. Using right-click menu options, you can access several **LTI Viewer** controls and options, including 1) **Plot Type** (changes the plot type), 2) **Systems** (selects or deselects any of the

261

models loaded in the **LTI Viewer**), 3) **Characteristics** (displays key response characteristics and parameters), 4) **Grid** (adds grids to your plot), and 5) **Properties** (opens the **Property Editor**, where you can customize plot attributes). In addition to right-click menus, all response plots include data markers. These allow you to scan the plot data, identify key data, and determine the source system for a given plot. The **LTI Viewer** has a tool bar that you can use to do the following: 1) open a new **LTI Viewer**, 2) **Print**, 3) **Zoom in**, and 4) **Zoom out**.

11.3   Using the Simulink® LTI Viewer

For our LTI model we will use the same transfer function that was used in the **TF.mdl** we developed in Chapter 10. We will use it to create a new model called **TFLTI.mdl**. If you were to generate this model from scratch, you would open the Simulink **Library Browser** using the MATLAB® command **simulink**. You would then select **File**, **New**, **Model**. Remember that **Ctrl+N** can be used as a shortcut to accomplish this. A new, untitled model window would be opened. You would then save this model under the name **TFLTI.mdl** using **File**, then **Save as. . .** as you have done in previous sections.

We will add three elements to our **TFLTI** model window. The model itself will require a **Transfer Function** block. You can either take the transfer function from the **Continuous** systems library as before or make a copy from your **TF.mdl** Simulink model. Here we will use the **Transfer Fcn** model from **TF.mdl** as was done in Chapter 10. This model will also require an LTI **Input Point** and an LTI **Output Point**. These Simulink model elements appear as follows.

Transfer Fcn block:

or   (TF.mdl)

Input and Output: These are used to designate the input and output signals to be retained in the linear approximation. In general, you choose signals that will be connected to a controller by right-clicking on a signal and selecting either an **Input Point** or an **Output Point** from the **Linearization Points** submenu.

If you take the **Transfer Fcn** model from the **Continuous** systems library as in Chapter 10, the transfer function properties will need to be modified using the **Block Parameters** window shown in Fig. 11.1.

Block Parameters: Transfer Fcn   ✕

Transfer Fcn

Matrix expression for numerator, vector expression for denominator.
Output width equals the number of rows in the numerator. Coefficients are
for descending powers of s.

Parameters
Numerator:

[2,1]

Denominator:

[2,5,3]

Absolute tolerance:

auto

OK   Cancel   Help   Apply

The approach we will follow in the remainder of this section is to reuse the
TF.mdl file we developed in Chapter 10. First delete the input **Sink** and
output **Source** elements from **TF.mdl**. It will then appear as in Fig. 11.2.

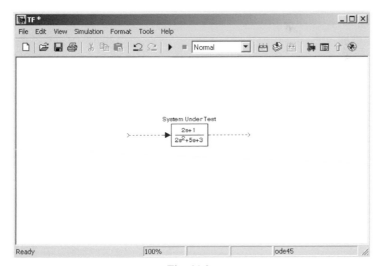

Fig. 11.2

The **TF.mdl** system for use in the **LTI Viewer** is next constructed by adding an LTI **Input Point** and an LTI **Output Point** and saving the model under the new name. As mentioned before, this is done to designate the input and output signals to be retained in the linear approximation. In general, you choose signals that will be connected to a controller by right-clicking on a signal and selecting either an **Input Point** or an **Output Point** from the **Linearization Points** submenu.

We will build our model using the **Control System Toolbox** library. You may find it desirable to build the transfer function using the **LTI System** block and then launching the **LTI Viewer** to obtain the LTI **Input Point** and the LTI **Output Point** (see Fig. 11.3).

Fig. 11.3

The **LTI System** block accepts both continuous and discrete LTI models as defined in the **Control System Toolbox**. Transfer function, state-space, and zero-pole-gain formats are all supported in this block. Figure 11.4 demonstrates this modeling and analysis process.

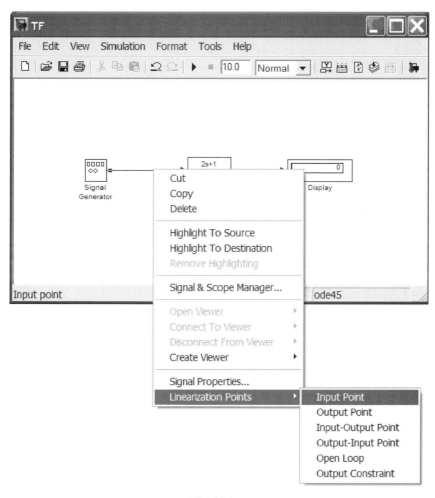

Fig. 11.4

Now select the **Input Point** from the **Linearization Points** on the right-click menu. If you place the **Input Point** close enough to the input arrow going into the **Transfer Fcn (System Under Test)** block, Simulink will automatically connect these two elements. If it does not, manually connect the elements using either the mouse drag method or the select and point shortcut covered in Chapters 9 and 10. After this is accomplished, your diagram will appear as in Fig. 11.5. Note the asterisk * showing the unsaved modification to the diagram. We will again save the model under the new name **TFLTI.mdl** after we have added the **Output Point**.

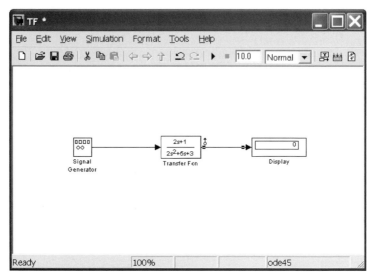

Fig. 11.5

After we have added the **Output Point**, the model appears as in Fig. 11.6.

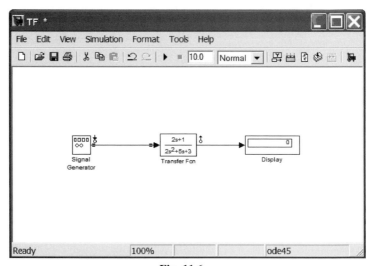

Fig. 11.6

Now we will save the model under the new name **TFLTI.mdl**. As a reminder, this model is saved under a new name first using **File**, **Save as. . . .** This process is repeated graphically in Fig. 11.7.

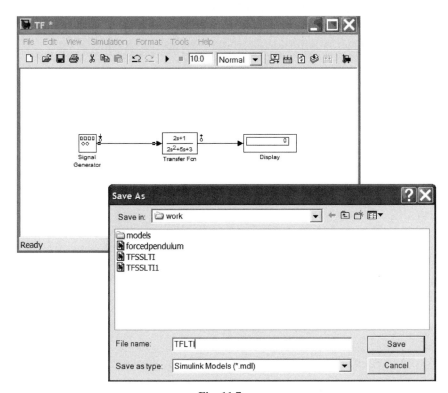

Fig. 11.7

We are now ready to launch the **LTI Viewer**. The **LTI Viewer** is opened by selecting **Tools, Control Design**, and **Linear Analysis. . . .** This command is executed as shown in Fig. 11.8. Note that this will take a few seconds and that you will be shown how close to completion the launch process is as illustrated.

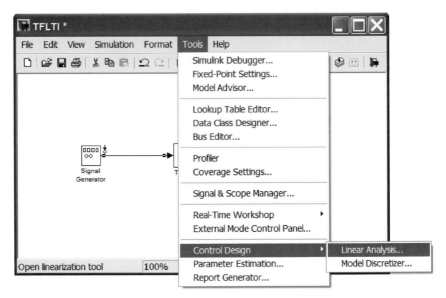

Fig. 11.8

The **LTI Viewer** initially appears as in Fig. 11.9. The default system response is a **Step Response**. You are initially given the opportunity to have the **LTI Viewer** open up the **Help** window to point specifically to its help documentation.

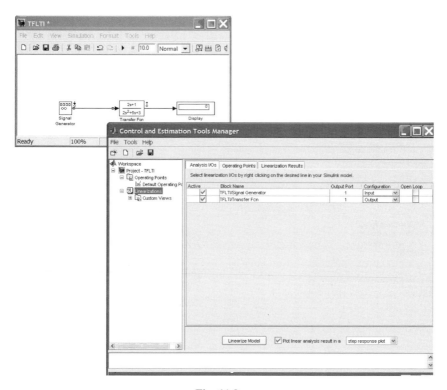

Fig. 11.9

The **LTI Viewer Help** is contained within the MATLAB **Help Navigator**. If you were to respond by selecting the **Help** button from the window in Fig. 11.9, the **Help** window in Fig. 11.10 would be opened.

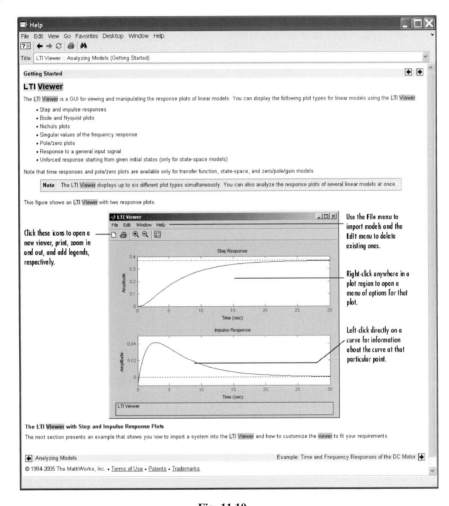

Fig. 11.10

Next we will use the variety of excitation and plotting methods available in the Simulink **LTI Viewer**.

Note that you will need to **Get** your linearized model (**TFLTI.mdl**) and load it into the **LTI Viewer**. Using the **LTI Viewer: TFLTI** window, select the **Simulink** pull-down menu and then select **Get Linearized Model**. The result of this operation will appear as in Fig. 11.11.

Note that because the default analysis option is set to **Step Response**, the time response at the **Output Point** due to a step input into the **Input Point** is then shown in the **LTI Viewer** window (Fig. 11.12). The plot scale is reset to the time that the response has reached its steady-state value.

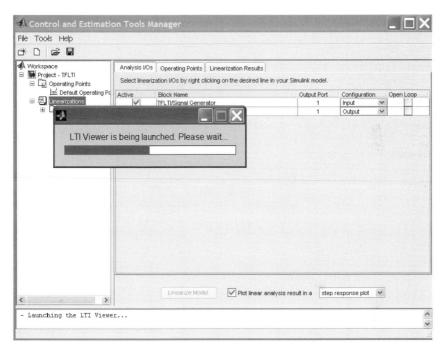

Fig. 11.11

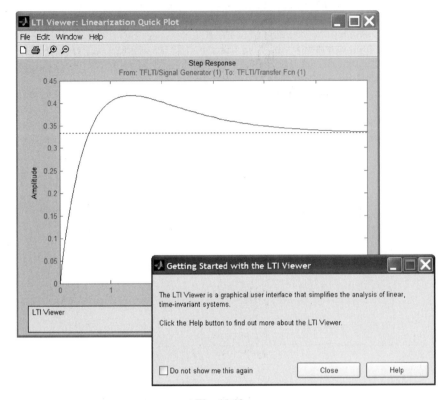

Fig. 11.12

It might be useful to refer back to Table 3.1 in Chapter 3 summarizing the equivalences between the different analysis methods available through the **LTI Viewer**. There they are listed for 12 different system types.

We will now examine each of these analysis approaches using the Simulink **LTI Viewer**.

There are two methods for selecting response plots in the **LTI Viewer**: 1) selecting **Plot Types** from the right-click menu and 2) opening the **Plot Configurations** window. The **Plot Configurations** window is listed under the **Edit** pull-down menu. If you have a plot open in the **LTI Viewer**, you can switch to any other response plot available by selecting **Plot Types** from the right-click menu. Figure 11.13 shows the right-click menu with **Plot Types** selected.

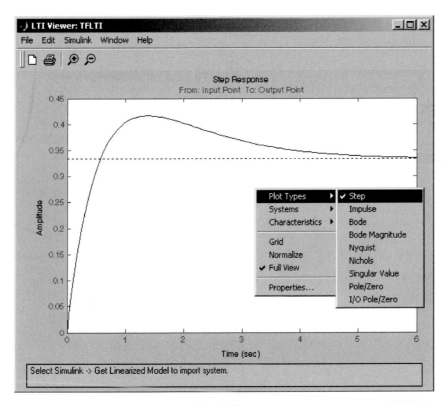

Fig. 11.13

To change the response plot, select the new plot type from the **Plot Types** submenu. The **LTI Viewer** automatically displays the new response plot.

To change the **Properties** of the plot you are viewing, select **Properties...** from the same right-click menu. The available **Properties** change from plot type to plot type. Figures 11.14–11.18 list all the **Properties** for the **Step Response**.

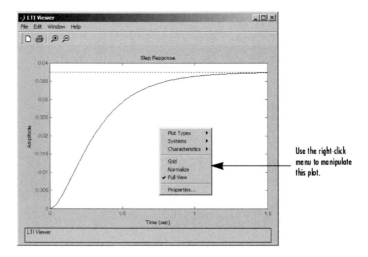

Fig. 11.14

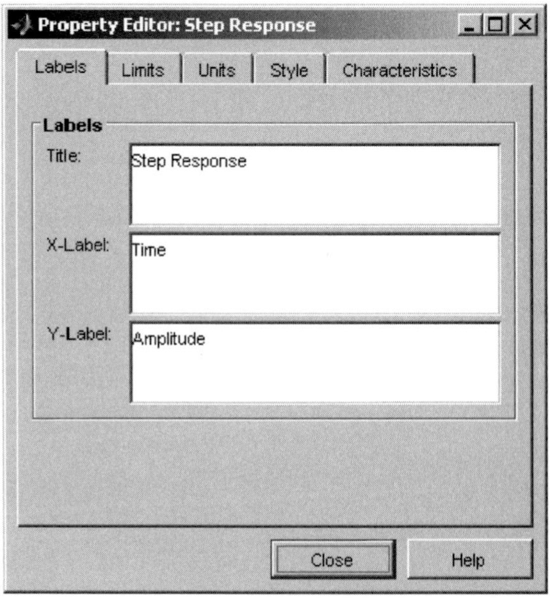

Fig. 11.15

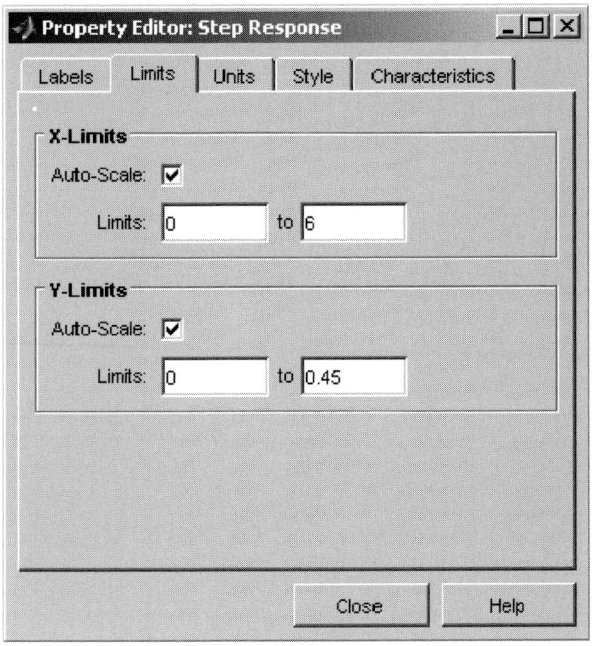

Fig. 11.16

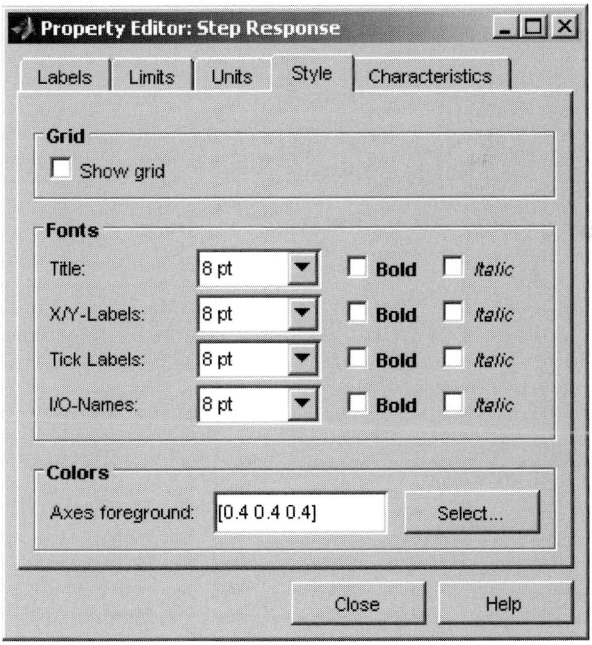

Fig. 11.17

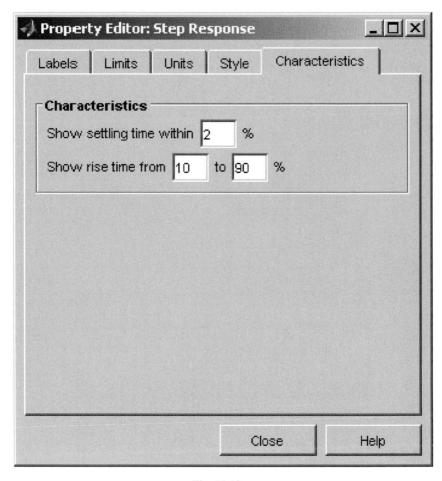

Fig. 11.18

All of the available plot types are shown in Figs 11.19–11.26 for our system **TFLTI.mdl**. Refer to Table 3.1 of 12 different transfer function responses as required. Note that some of the plots have special grids and other plotting options available for that particular analysis type.

Impulse Response:

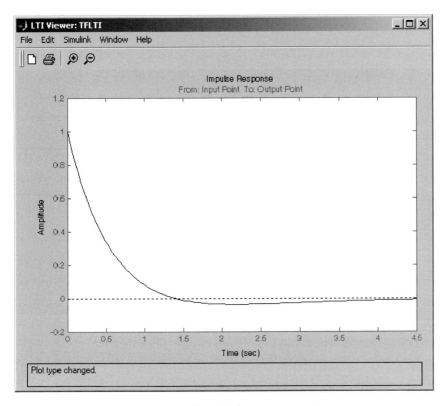

Fig. 11.19

Bode Diagram (magnitude and phase):

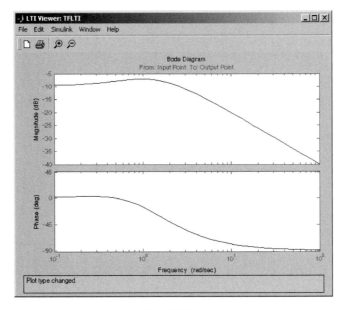

Fig. 11.20

Bode Diagram (magnitude):

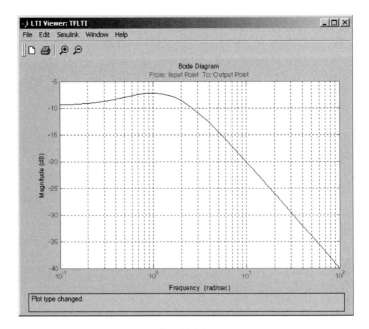

Fig. 11.21

Nyquist Charts:

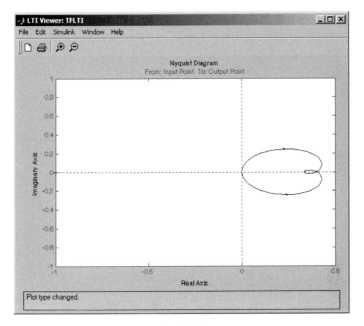

Fig. 11.22

Nichols Charts:

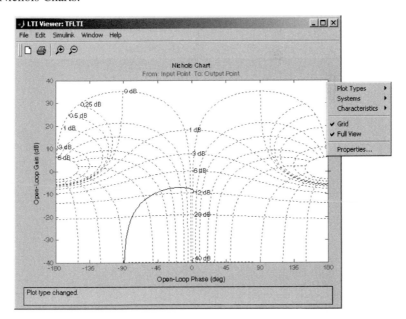

Fig. 11.23

Sigma Plot:

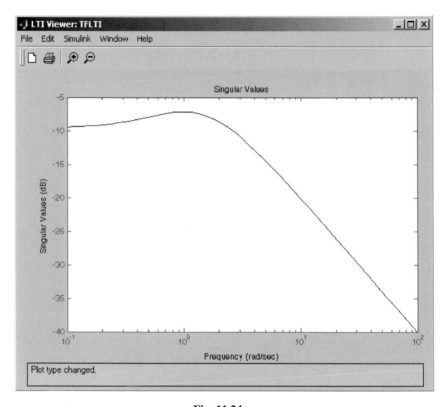

Fig. 11.24

Pole/Zero Plots:

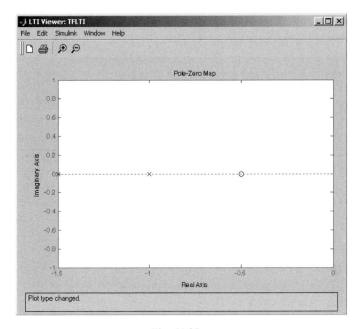

Fig. 11.25

I/O Pole/Zero Plots:

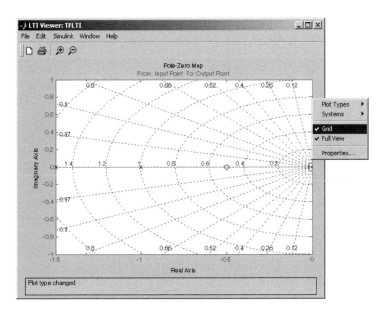

Fig. 11.26

The **Plot Types** feature of the right-click menu works on existing plots, but you can also add plots to an **LTI Viewer** by using the **Plot Configurations** window.

To reconfigure an open viewer, select **Plot Configurations...** under the **Edit** menu. This opens the **Plot Configurations** window. This action and the resulting **Plot Configurations** window are shown in Figs 11.27 and 11.28.

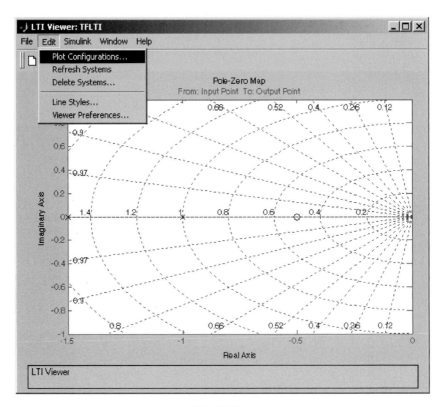

Fig. 11.27

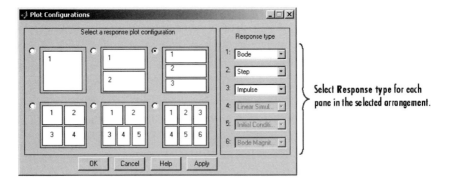

Fig. 11.28

11.4 Equivalent Simulink® LTI Models

In Chapter 10 we generated equivalent **Zero-Pole** and **State-Space** models to the transfer function model we have been studying:

$$\frac{2s + 1}{2s^2 + 5s + 3}$$

These equivalent models are as follows. The equivalent **Zero-Pole** model is

$$\frac{s + 0.5}{(s + 1.5)(s + 1.0)}$$

Replacing the **Transfer Fcn** block in the model **TFLTI.mdl** with the **Zero-Pole** model results in the model shown in Fig. 11.29. Note that it gives exactly the same results in the **LTI Viewer** as are seen in Sec. 11.3.

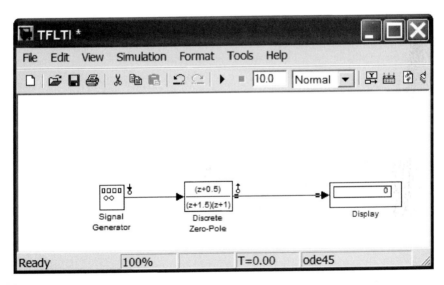

Fig. 11.29

The equivalent **State-Space** model is

$$\dot{x} = Ax + Bu$$
$$y = Cx + Du$$

where

A =
   −2.5000   −1.5000
     1.0000   0
B =
    1
    0
C =
    1.0000   0.5000
D =
    0

Replacing the **Transfer Fcn** block in the model **TFLTI.mdl** with the **State-Space** model results in the model shown in Fig. 11.30. Note that it gives exactly the same results in the **LTI Viewer** as are seen in Sec. 11.3.

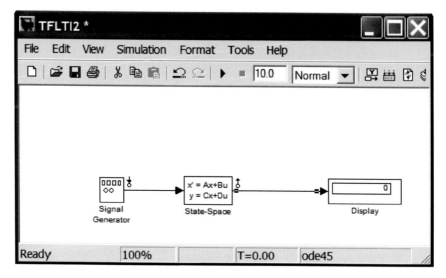

Fig. 11.30

A specific block for the **LTI Viewer** is also available. It is found in the **Control System Toolbox** library. The **SISO Design Tool** discussed in Sec. 11.5 also uses this block (see Fig. 11.31).

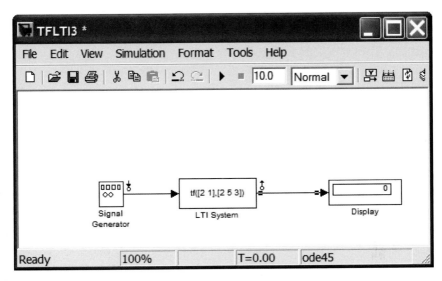

Fig. 11.31

11.5 SISO Design Tool

The **SISO Design Tool** is a Simulink graphical user interface that allows you to analyze and tune SISO feedback control systems. Using the **SISO Design Tool**, you can graphically tune the gains and dynamics of a compensator and a prefilter using root locus and loop-shaping techniques.

Using the **SISO Design Tool**, you can use the root locus view to stabilize the feedback loop and enforce some minimum damping. You can also use the Bode diagrams to adjust the bandwidth, check the gain and phase margins, or add a notch filter for disturbance rejection. You can also generate an open-loop Nichols view or Bode diagram of the prefilter by selecting these items from the **View** menu. All views are dynamically linked. Changing any parameter or element, such as the gain in the root locus, will immediately update the Bode diagram.

The **SISO Design Tool** is designed to work closely with the **LTI Viewer**, allowing you to rapidly iterate on your design and immediately see the results in the **LTI Viewer**. When you make a change in your compensator, the **LTI Viewer** associated with your **SISO Design Tool** automatically updates the response plots that you have chosen. By default, the **SISO Design Tool** displays the root locus and open-loop Bode diagrams for your imported systems. You can also generate an open-loop Nichols view or prefilter Bode diagram by selecting these items in the **View** menu. Imported systems can include any of the elements of the feedback structure diagram located to the right of the **Current**

Compensator panel. You cannot change imported plant or sensor models, but you can use the **SISO Design Tool** for designing a new (or for modifying an existing) prefilter or compensator for your imported plant and sensor configuration. The default **SISO Design Tool** window appears in Fig. 11.32.

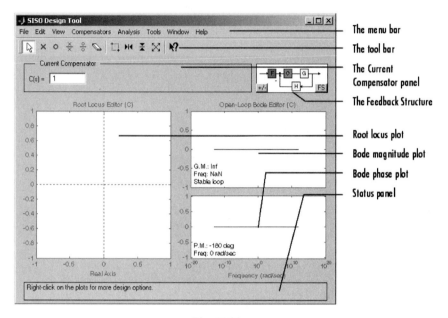

Fig. 11.32

Start the **SISO Design Tool** from the MATLAB **Command Window** by typing the command **sisotool**. You will be informed on the progress of the tool launch sequence. You will then be pointed to the MATLAB **Help** window on this tool upon your request (see Fig. 11.33).

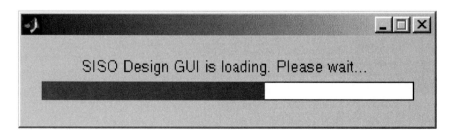

Fig. 11.33

The tool window will look as in Fig. 11.34.

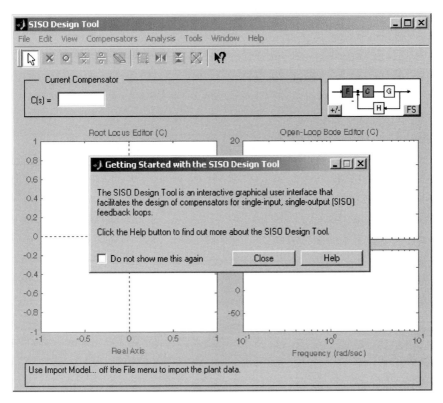

Fig. 11.34

You can generate the model for our transfer function model for use in the **SISO Design Tool** in the MATLAB **Command Window** by typing the following:

```
>> num = [2, 1]

num =
    2   1


>> den = [2, 5, 3]

den =
    2   5   3

>> sisosys = tf(num,den)
```

The resulting transfer function is of the form

$$\frac{2s + 1}{2s^2 + 5s + 3}$$

Next use **File**, **Import. . .** to load the model into the tool (see Fig. 11.35).

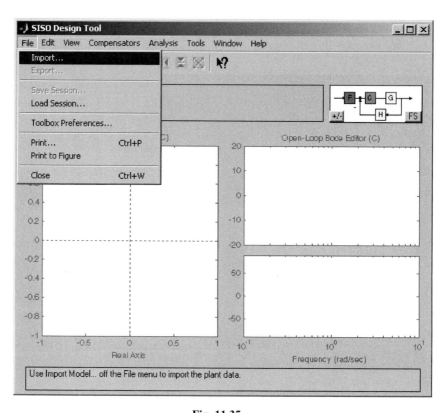

Fig. 11.35

Note that **sisosys** is already visible under **SISO Models** (see Fig. 11.36).

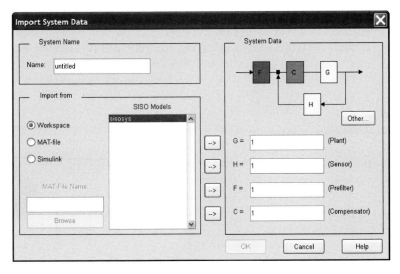

Fig. 11.36

Load **sisosys** into the **C** block, and set the value of **F** to near zero. Note that the **SISO Design Tool** does not allow you to set a block to exactly zero (you will get a warning message if you attempt this). Save the model under the name **tfsys**. Your **Import System Data** window will look as in Fig. 11.37.

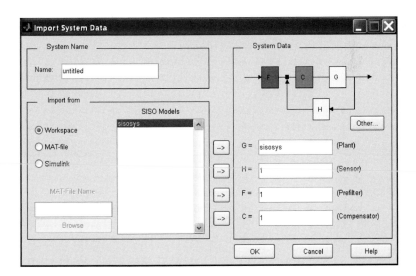

Fig. 11.37

When you select OK in the **Import System Data** window, your **SISO Design Tool** window will look as in Fig. 11.38.

Note that these responses are the same as those we saw using the **LTI Viewer**. The **x** and **o** denote the open-loop poles and zeros, respectively. The squares denote the roots of the unity feedback system. The response of the system can now be modified as desired by adjusting the gain value or adding pre- or postfilters or system feedback. You can also interact with the system response by adjusting the pole and zero locations. Try out the different design features and model configurations (use the **FS** radio button). An example of this process is shown in Fig. 11.38.

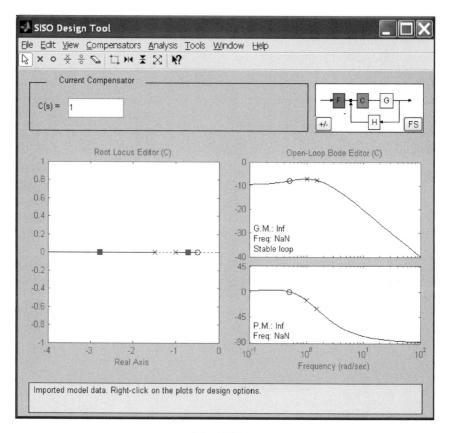

Fig. 11.38

We could also have loaded our system into the **SISO Design Tool** using the **LTI System** block. Try this out yourself. Make a simple system called **test.mdl** by creating a Simulink model using an **LTI System** block. Enter the same numerator and denominator that we just studied using the **SISO Design**

Tool. You need to open the **LTI System Block Parameters** by double-clicking on the block element. The **Block Parameters** window after modification will look as in Fig. 11.39.

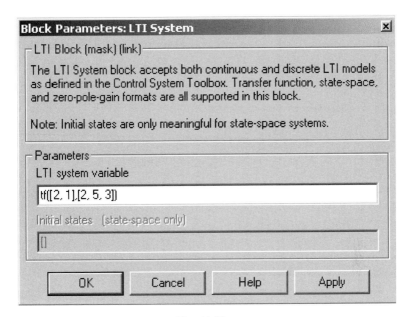

Fig. 11.39

After selecting **OK**, this Simulink model window will look as in Fig. 11.40.

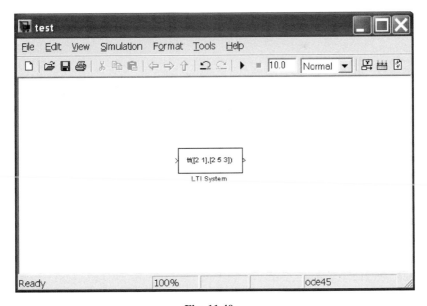

Fig. 11.40

Now use the **File, Import...** option to load our new **LTI System** block into the **SISO Design Tool**. Select the **Import from, Simulink** option into the model we call **tfsys**. If the model does not appear in the window explicitly, select the **Browse** option to locate it (see Fig. 11.41).

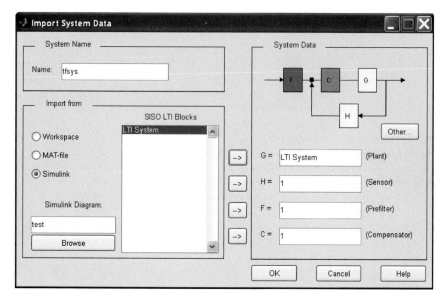

Fig. 11.41

If the model is open, you may get the warning shown in Fig. 11.42.

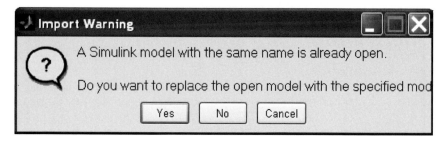

Fig. 11.42

It does not matter if you select **Yes** or **No**, although the safe selection is **No**. Select the **LTI System** from the list of **SISO LTI Blocks** and replace the previous compensator with the test as was shown in the **Import System Data** window (Fig. 11.41).

Your system's response should appear unchanged. You might want to select **File, Save Session** for future reference as shown in Fig. 11.43.

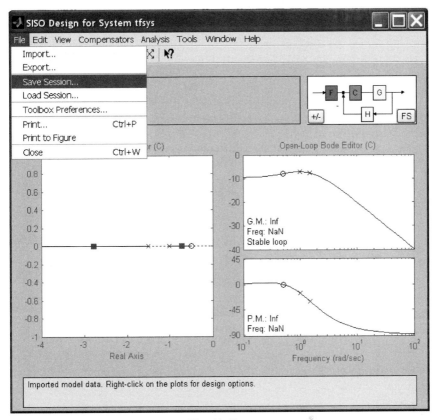

Fig. 11.43

11.6 Conclusion

This chapter was an introduction to using the Simulink **LTI Viewer** for the analysis of linear, time-invariant dynamic systems. It was also a very brief introduction to using the **SISO Design Tool** for the design of LTI systems. Further examples are given in the Simulink **Help** window. Some of these examples demonstrate the use of these tools to build complex multiple-input, multiple-output systems.

Practice Exercises

11.1   This problem is to put the following transfer function from Chapter 10 into **Zero-Pole** form and then analyze it using the **LTI Viewer**. That transfer function is

$$\frac{s^2 + 4s + 3}{2s^2 + 5s + 2}$$

In Chapter 10 you used the MATLAB **tf2zp** command to convert this into **Zero-Pole** form and implement this within Simulink. Now replace the inputs and outputs with those for the **LTI Viewer** as in Sec. 11.3. Next replace the **Zero-Pole** elements in your model with the equivalent **State-Space** element. In the previous exercise, you used the MATLAB command **tf2ss** to do this. Verify that the responses are the same.

For each of these two system representations generate the eight different plot types: 1) step response, 2) impulse response, 3) Bode diagrams (either magnitude and phase or just magnitude), 4) Nyquist charts, 5) Nichols charts, 6) sigma plot, 7) pole/zero plots, and 8) I/O pole/zero plots.

Notes

12
Building a Multiple-Input, Multiple-Output Simulink® Model

12.1 Introduction and Objectives

This chapter will introduce the reader to multiple-input, multiple-output systems implemented in Simulink®. Its dynamic modeling capabilities will also be demonstrated.

Upon completion of this chapter, the reader will be able to build dynamic multiple-input, multiple-output Simulink models using Simulink elements; use **Mux/Demux** and **Bus** methods for generating and using vector signals; generate subsystem models; use masking; generate libraries; execute and simulate multiple-input, multiple-output systems implemented in simulink, and directly interface with these models via MATLAB®.

12.2 System Modeling in Simulink®

Using Simulink, complex multiple-input, multiple-output systems can be modeled and simulated. A variety of different elements are contained within the Simulink libraries and can be easily interconnected using Simulink. A variety of inputs can be examined simultaneously using a large number of different output devices. These devices can display this information in either graphical or numeric form. These data can be made available to MATLAB for further analysis. Data can also be accessed by the model from MATLAB. The model itself can be accessed and manipulated from the MATLAB **Command Window** or using M-files and MEX-files.

12.3 Parameter Estimation

We will now build a multiple-input, multiple-output system for parameter estimation. We will duplicate the simple **TF** system that we generated in Chapter 10, and then we will connect these two systems together for parameter estimation. Other elements and connections will be added to complete the desired system. We will call this new system **parmest.mdl**.

This system will be used to estimate a single scalar parameter. It will take the difference between the output of the system under test and the output of a system model using the estimated parameter to generate an error signal. The integral of the product of this error signal and the system input is the estimate of the unknown parameter. The value of the gain in the block labeled **Parameter** is the parameter to be estimated.

One way to think of this is as an experiment to tune a mathematical model of a system based on the input-output behavior of the system undergoing testing. The MathWorks provides MATLAB and Simulink tools for such real-time modeling and data acquisition.

The system we will build looks as follows. We have two **Signal Generators**, our two **Transfer Fcn** systems, and three different **Scopes**. A variety of other interconnecting elements are contained within this model.

In this section we will go through the process of generating this system model in Simulink. In accomplishing this, we will take advantage of the model that we developed and analyzed in Chapters 10 and 11 (see Fig. 12.1).

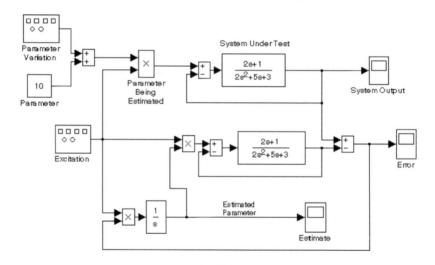

Fig. 12.1

First we need to open up a new Simulink model and give it the name **parmest.mdl**. Do this by opening the **Simulink Library Browser** using the MATLAB command **simulink**. Then select **File**, **New**, and **Model**.

Note that **Ctrl + N** can be used as a shortcut to accomplish this action. A new, untitled model window is opened. Save this model under the name **parmest.mdl** using **File**, then **Save as . . .** You now have a new Simulink model window in which to work.

We will now take advantage of the single-input, single output system model **TF.mdl** that you generated in Chapter 10. You can have multiple Simulink models open at the same time. Use the **Simulink Library Browser** to open up this model. You can also use **Ctrl + O** as a shortcut to access this model. Next, go back to the **TF** model window. Select your model from the window by using the right mouse button while sweeping the mouse across all three of the model elements. This will select the model for copying. When you release the mouse button, you can either use the **Ctrl + C** shortcut to place a copy of this model into memory, or you can use the **File**, **Copy** option at the top of the **TF** window (see Fig. 12.2).

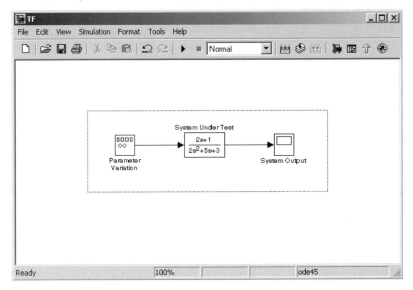

Fig. 12.2

Once you have selected the model, you can paste your model from **TF.mdl** into **parmest.mdl** twice. Your **parmest** model will now look like the windows shown in Fig. 12.3.

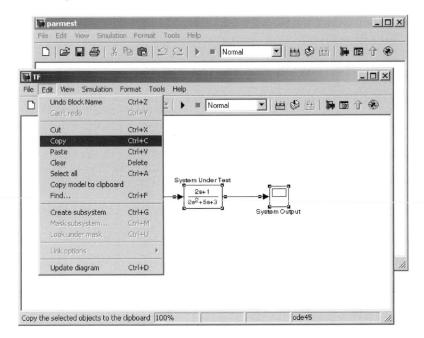

Fig. 12.3

After pasting the model twice, your **parmest** model appears as in Fig. 12.4.

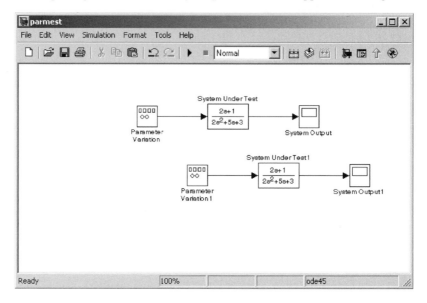

Fig. 12.4

We next need to add four summation junctions (**Sum**), three multipliers (**Product**), one scalar input parameter (**Constant**), and one additional **Scope** to our model. This will be demonstrated next.

The following libraries are used to get these elements.

First, from the library of **Math Operations** we get the summation junctions and the multipliers. Note that the default is a circular summation junction. We will need to change it into a rectangle. Modifying that parameter in the summation junction's **Parameter Window** does this.

We get the integrator from the set of **Continuous** blocks. To show that multiple elements can be used to model the same dynamic system, we will use a **Transfer Fcn** and build an integrator from it.

We will get our constant input parameter of **10** from the library of **Sources**.

Finally, we will get our last scope from our library of **Sinks**. We could also select one of the **Scopes** already in the **parmest** model and copy and paste it into the model as a shortcut.

To construct the model, first we paste a summation junction (**Sum**) into our model and then modify its parameters list to make it rectangular. The result of this action is shown in Fig. 12.5.

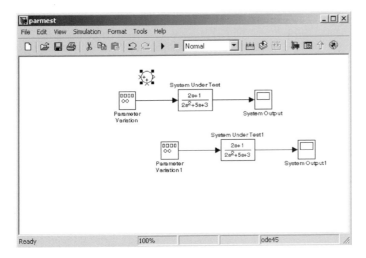

Fig. 12.5

Note that this is a circular summation junction. We change it into a rectangle using the summation junction's **Parameter Window** as shown in Fig. 12.6.

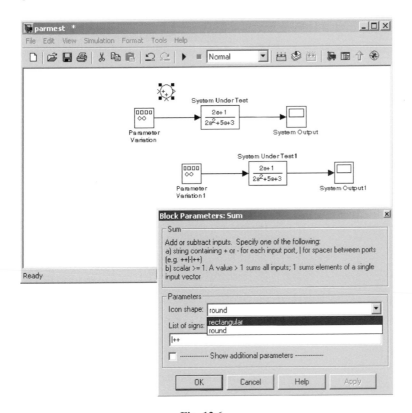

Fig. 12.6

Next we make multiple copies of this summation junction (**Sum**) and paste them as needed into the diagram. We need to change some of the signs as appropriate to match our desired model, using the same **Block Parameters** window. After we have completed these actions, our model appears as in Fig. 12.7.

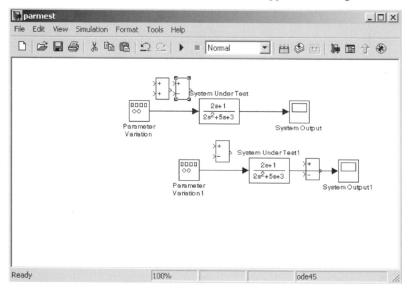

Fig. 12.7

Next we add the **Product** block to our diagram.
The diagram with all of the **Product** blocks added is shown in in Fig. 12.8.

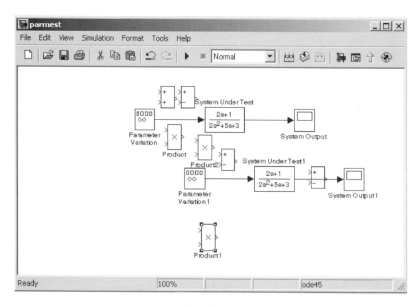

Fig. 12.8

Our diagram is getting a little crowded. To get enough room to paste in the remaining elements, we must select and drag the elements around in the window to approximate the final spacing.

The result of the actions to make room for the remaining block elements is shown in Fig. 12.9.

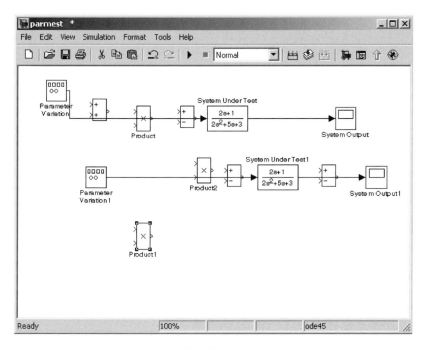

Fig. 12.9

Now we have the space to add our remaining elements to our model window. First we need to add the final input parameter. This is a **Constant** from the **Sources** library. The default value for this bias is **1**.

Most block elements come with default values or settings. We will need to change the **Constant value** later to **10**, just as we modified our **Sum** properties after placing them into our system diagram. We next select another **Transfer Fcn** from the **Continuous** library and paste it in (see Figs 12.10 and 12.11).

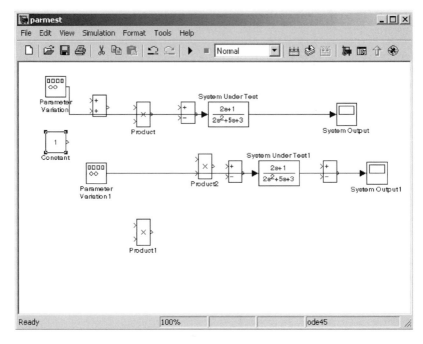

Fig. 12.10

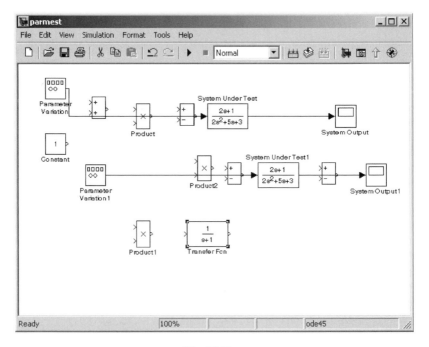

Fig. 12.11

Now select a **Scope** from the **Sinks** library and drop it at the desired location into our model window (Fig. 12.12).

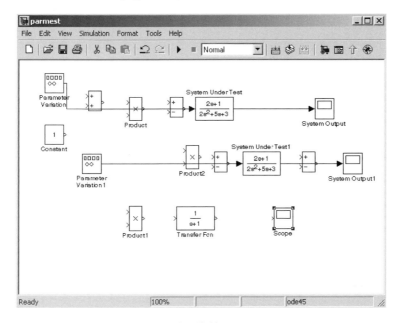

Fig. 12.12

Double-click on the **Constant** block, just as we did when we modified the **Sum** properties after placing them into our system diagram. Change the **Constant value** to **10** (Fig. 12.13).

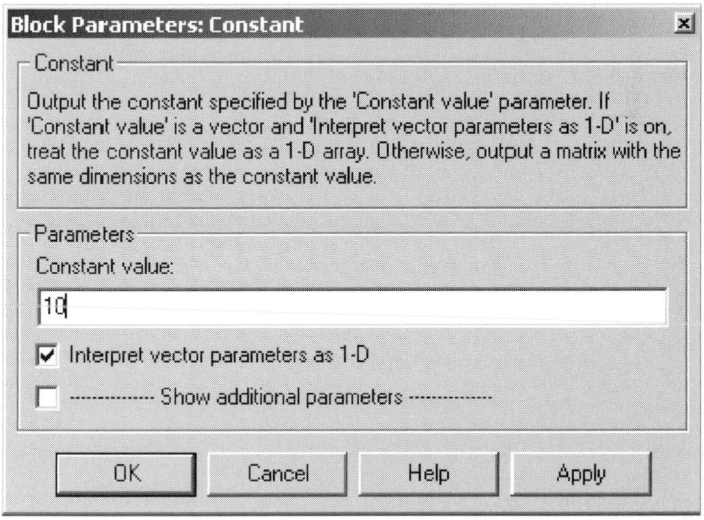

Fig. 12.13

The value changes to 10 on the element after **Apply** is selected (see Fig. 12.14).

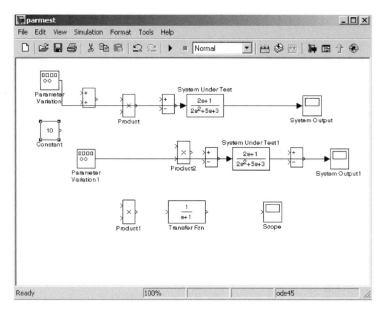

Fig. 12.14

Now double-click on the **Transfer Fcn** and modify its **Parameters** as shown in Fig. 12.15.

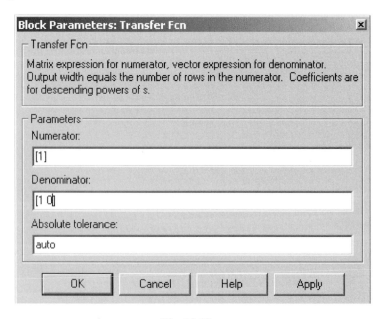

Fig. 12.15

Again, when **Apply** is selected, the **Apply** text will turn gray and the new **Transfer Fcn** model will appear on the block in the **parmest** diagram as shown in Fig. 12.16.

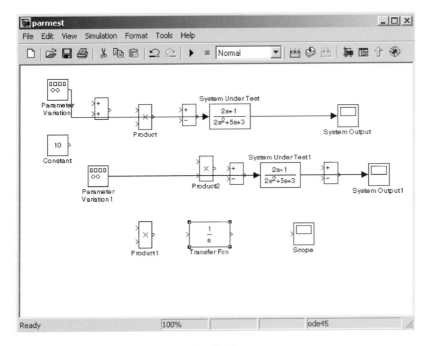

Fig. 12.16

We have now successfully added all the elements we need into our **parmest** diagram. Now we need to connect the blocks. We start with the **Parameter Variation** input and connect it to the summation element. I first use the drag and release method. Simulink reminds me that a shortcut for this operation is available (see Fig. 12.17).

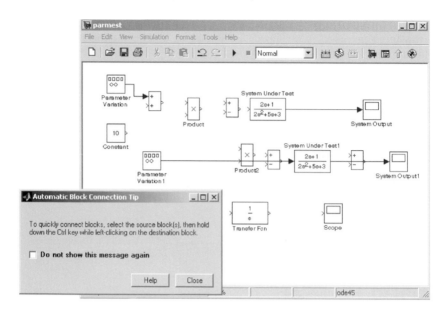

Fig. 12.17

We can repeatedly use this shortcut to connect all the elements across the top of our **parmest** Simulink diagram. When these actions are repeated and the model saved, the model window will appear as follows (see Fig. 12.18).

Also shown next is the remainder of the connections that do not require a signal to branch (see Fig. 12.19). Branches are selected off of the line elements in a similar way to how the block elements are created. Just right-click onto the line where the branch is desired and drag to the desired input location. This process is demonstrated in Figs 12.20 and 12.21 that follow. Multiples of these branching connections are required to complete this system diagram.

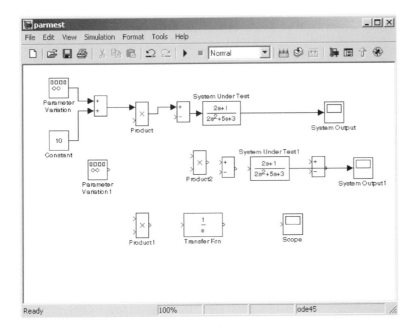

Fig. 12.18

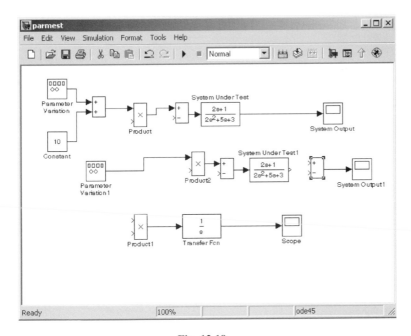

Fig. 12.19

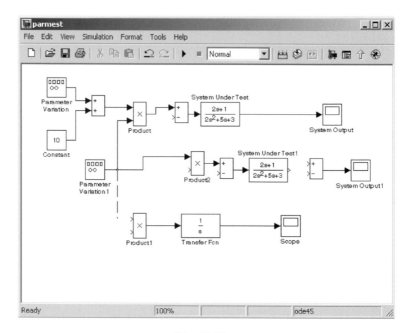

Fig. 12.20

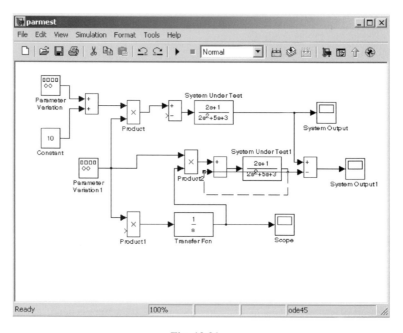

Fig. 12.21

Note that Simulink does not necessarily route these element-to-element connections in the way that you desire. You can select and move the connections and change their routings in the same way that you move other Simulink elements. In Fig. 12.21, the bottom output of the **System Under Test1** element was selected. This output is then connected in a feedback loop to the bottom input of the previous **Sum**. Simulink automatically draws the feedback path over the top of the **Transfer Fcn** block. Just select this element and drag it below the bottom of the **Transfer Fcn** block as desired. Corners of connection routings can also be similarly moved, and additional segments added to routings. If Simulink can replace a multiple segment routing with a straight segment, it will automatically do so (see Fig. 12.22).

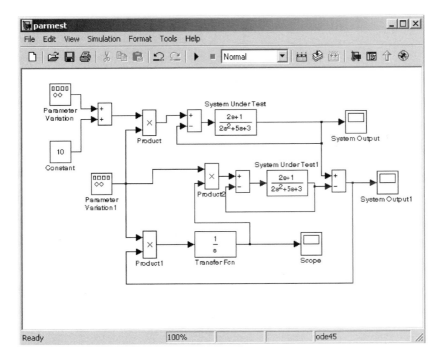

Fig. 12.22

The block element names need to be modified to match those desired names given at the beginning of this section. Some of the names that are not required need to be deleted from the diagram. Select the desired label. If no name is required, the name field can be deleted using the **Delete** key. Note that no Simulink element can have the same name. Simulink will add a number after the name for a repeated element. Simply deleting the characters from a name will not remove the name field. Two blank text fields will cause Simulink to add a **1** to the second blank field. The next figure shows the third **Scope** being given the name **Estimate** (see Fig. 12.23).

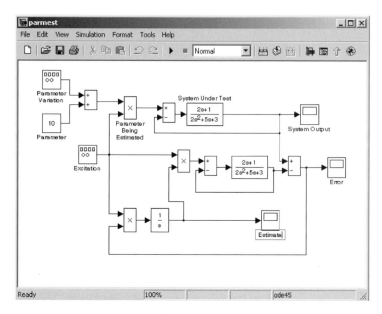

Fig. 12.23

The **Excitation Signal Generator** next needs to be modified as shown in Figs 12.24 and 12.25.

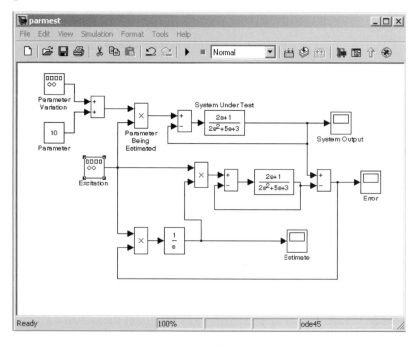

Fig. 12.24

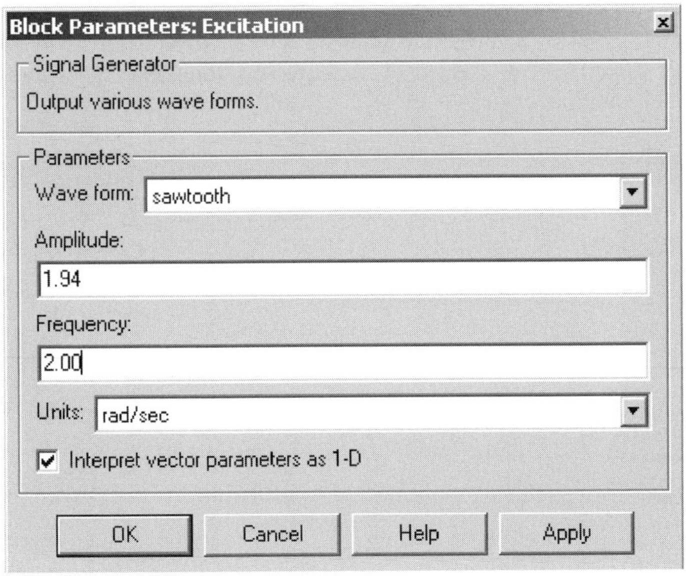

Fig. 12.25

The **Parameter Variation Signal Generator** is similarly modified as shown in Fig. 12.26.

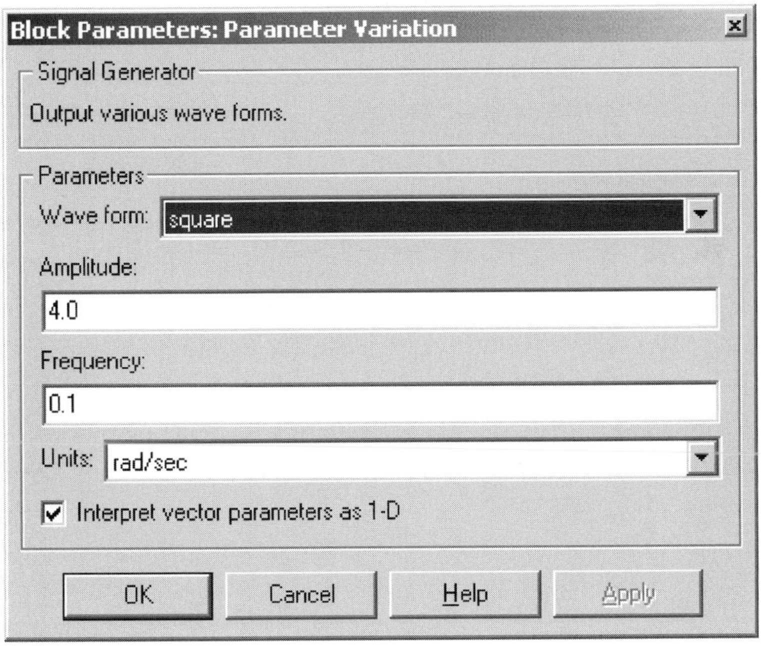

Fig. 12.26

Finally, the Simulation Parameters for **parmest.mdl** need to be modified. Normally both the **Solver** and the **Workspace I/O** need to be modified to the values you select. Here I have just added a very large **Stop time** so that the model can be run through multiple cycles (see Figs 12.27 and 12.28).

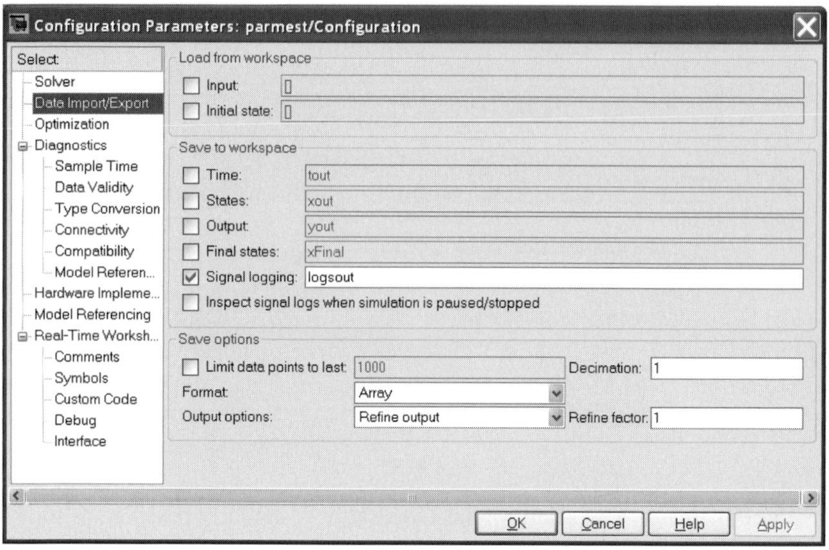

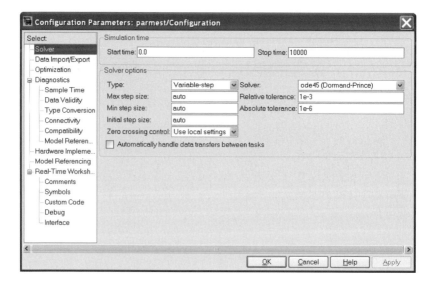

Fig. 12.27

Fig. 12.28

To see what is going on, open (double-click) and position the scopes. Then select **Simulate, Start** from the pull-down menu. You will see the time histories of the system response. The response should be as shown in Fig. 12.29.

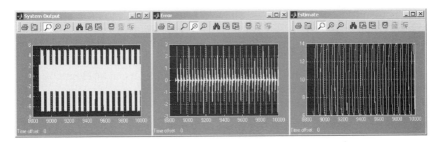

Fig. 12.29

Use the binocular symbol to autoscale the data and the magnifying glasses to zoom into and out of the figure. Try different finish times.

12.4   MATLAB® Simulation Interface

After you have tried different options to modify simulation and display parameters for your **parmest** model in Simulink, let us modify them using the MATLAB **Command Window**. First, let us look at one convergence of the estimated model parameter by overriding the simulation end time from MATLAB.

The **parmest** system can be simulated using the command line

$\gg$ [t,x] = sim('parmest',[0,50]);

The output in the **Error** and **Estimate Scope** windows will appear as shown in Fig. 12.30.

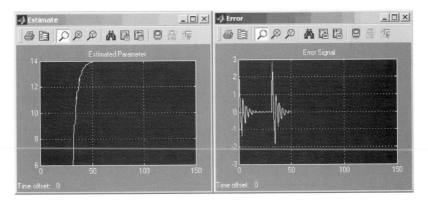

Fig. 12.30

Notice that as the estimate converges on the correct result, the error signal drops to zero.

If you select (click on) the **parmest** window, you can start and stop the simulation of this system yourself from the **Simulation** menu.

You may find it interesting to change the excitation (open it) and restart the simulation. Changing the gain in the **Parameter** block while the simulation is running also produces interesting behavior. The Simulink model is closed from MATLAB using the command

close_system('parmest',0);

The system can also be closed from the **File** menu, or it can be closed by clicking on the **x** at the upper right corner of the **parmest** model window.

12.5 Subsystems, Masking, and Libraries

Subsystems can be very useful in complex multi-input, multi-output models. They allow you to create multilevel models. New Simulink blocks can be constructed in this way. The models can even be masked, providing custom interfaces for these blocks. Finally, new libraries of these elements can be created.

Let us work with the model we built in Sec. 12.3. Some of the elements have been rearranged slightly to make the process of creating a subsystem a little easier. Also added are boxes to output data to the MATLAB workspace and to a **mat** file (see Fig. 12.31).

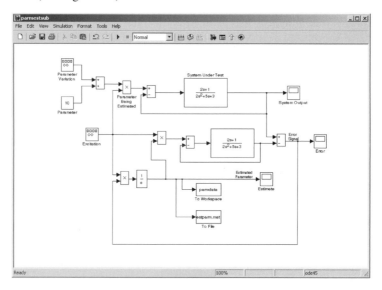

Fig. 12.31

To create a subsystem, use the bounding box. Select the elements you want to place into your subsystem. Note that there is no inverse to this operation, and so you might want to save before doing this operation. You might also need to rearrange your diagram so that you can easily select the desired elements (see Fig. 12.32).

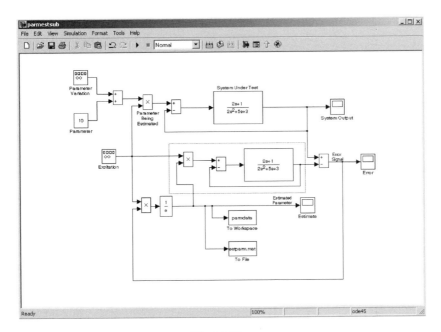

Fig. 12.32

Once the elements are selected, use **Edit, Create subsystem** (see Fig. 12.33).

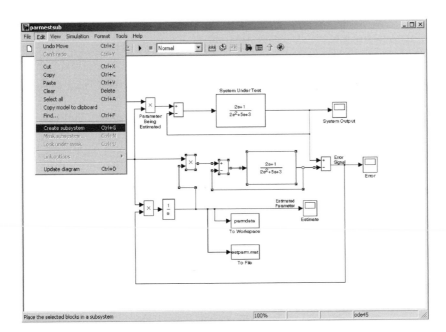

Fig. 12.33

The diagram now contains a subsystem as shown in Fig. 12.34.

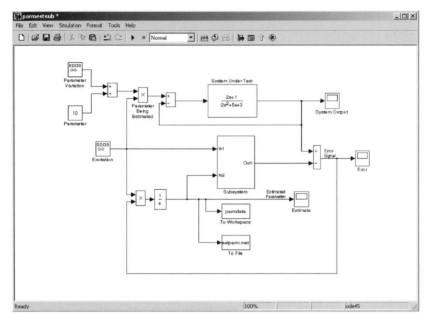

Fig. 12.34

Once the subsystem is created, all you need to do to open it is to doubleclick on the subsystem block. The subsystem window opens as shown in Fig. 12.35.

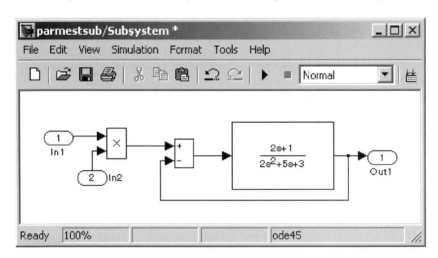

Fig. 12.35

Note that the default can be changed to open the subsystem in your existing model window rather than in a new window.

If the model is masked, the subsystem model will not be shown. Rather a **Block Parameters** window that you have built will be displayed.

In a following window a mask is generated for this model. First select the **Subsystem** block. With the selected block use **Edit, Mask subsystem . . . .** A **Mask editor** will then be evoked. Change the block **Icon, Parameters, Initialization**, and **Documentation** as desired.

In our example, the **Documentation** for this block is modified as shown in Figs 12.36 and 12.37.

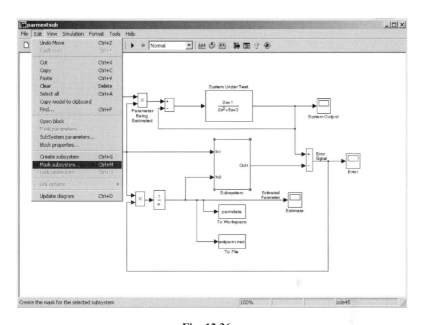

Fig. 12.36

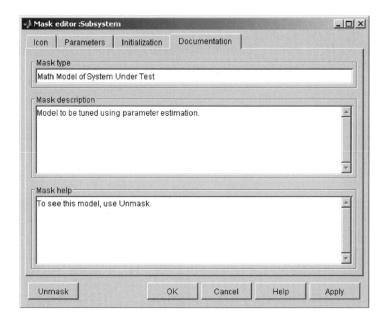

Fig. 12.37

After the aforementioned changes have been applied, the subsystem model will be masked. Double-clicking on the block will bring up the shown in Fig. 12.38.

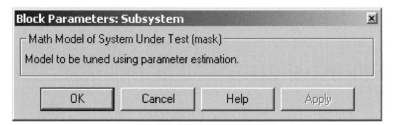

Fig. 12.38

To unmask the system, the **Unmask** option must be selected from the **Mask editor** window.

We can now create a new **Simulink Library** for our subsystem block. Note that one or many blocks can be added to our library once it is created. This is done in a manner nearly identical to the way we create Simulink models. The libraries even have the same **.mdl** extension reserved for Simulink models.

To create our **Simulink Library** go to the **Simulink Library Browser** and select **File**, **New**, **Library**. Note that there is not a keyboard shortcut for this operation.

Once this is done, you need to give a name to your library. You also need to save it.

We will create a library called **MyLibrary.mdl**. Here it is saved under **cdrive**. A copy of the subsystem is then saved in the **MyLibrary** window. The results of these actions are shown in Figs 12.39–12.41.

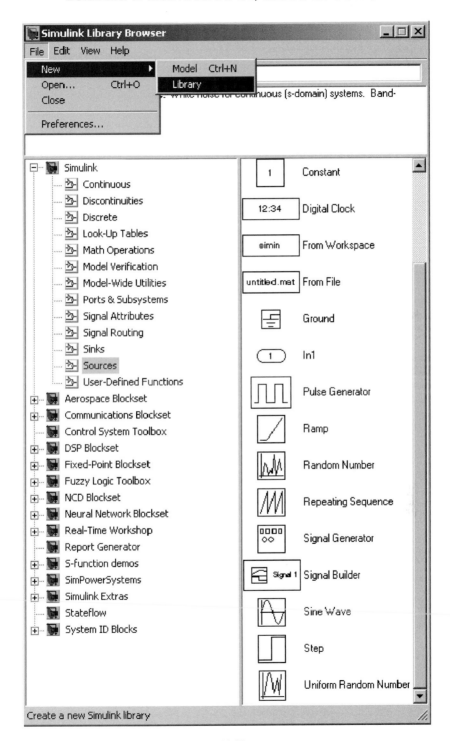

Fig. 12.39

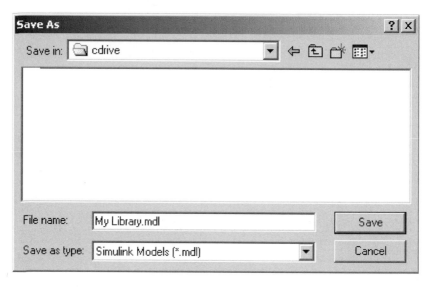

Fig. 12.40

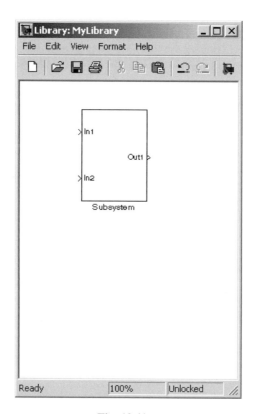

Fig. 12.41

12.6   Vector Signals

When modeling a multiple-input, multiple-output system, you can end up with a complex and confusing model if you only use scalar signals. Vector signals can be used to clean up these signal lines. There are two ways to accomplish this: 1) **Mux/Demux** and 2) **Bus Creator/Bus Selector**. These are found in the **Simulink Library Browser** under **Signal Routing**. **Bus Creator/Bus Selector** might be preferable to **Mux/Demux** if you are tracking signal names. The following diagrams show the differences between **Mux/Demux** and **Bus Creator/Bus Selector**.

First, three signals are fed into a **Mux** element. The **Mux** element generates a vector containing three signals. These three signals are converted back into scalar signals using the **Demux** element (see Fig. 12.42).

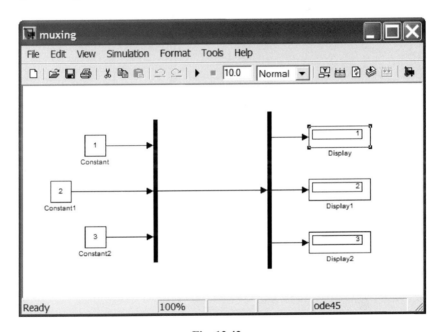

Fig. 12.42

In Fig. 12.43, three signals are fed into a **Bus Creator**. The **Bus Creator** element generates a vector containing three signals. The three signals have been named **a**, **b**, and **c**. These three signals are converted back to scalar signals using the **Bus Selector** element. The **Bus Selector** has kept track of the names of the signals on the bus.

It displays the names of the scalar signals as **<a>** , **<b>** , and **<b>** . This can be very useful when generating large, multilayer block diagrams.

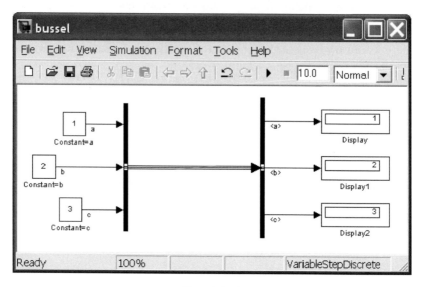

Fig. 12.43

Note that you can eliminate a diagnostic warning by setting the "**Automatic solver parameter selection**" diagnostic to "**none**" in the **Diagnostics** page of the **configuration parameters** dialog. Also note that **Format, Wide Nonscalar Lines** can be very useful when attempting to differentiate between scalar and vector signals. **Format, Vector Line Widths** is very useful because it shows you the number of separate signals on the vector connection (see Fig. 12.44). Try these out and see what you think!

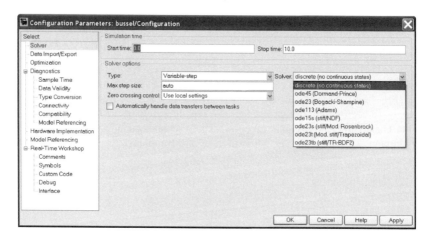

Fig. 12.44

12.7   Using Vector Signals for Math Functions

This section demonstrates the use of **Math Operations** and vector signals to accomplish within Simulink some of the mathematical operations demonstrated within MATLAB in Chapter 1. The **Math Operations** used are 1) **Adjoint of 3×3 Matrix**, 2) **Create 3×3 Matrix**, 3) **Determinant of 3×3 Matrix**, and 4) **Invert 3×3 Matrix**. These are all found within the **Aerospace Blockset** as in Fig. 12.45.

The model to accomplish this appears as follows. **Constant** blocks are used to enter the matrix elements. The block is used to construct the actual 3×3 matrix. The adjoint, determinant, and matrix inverse are then calculated using the appropriate blocks. **Display** blocks are used to show the results (see Fig. 12.46).

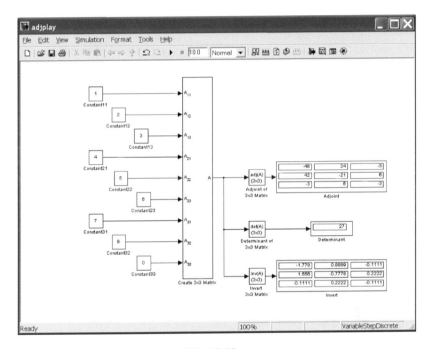

Fig. 12.45

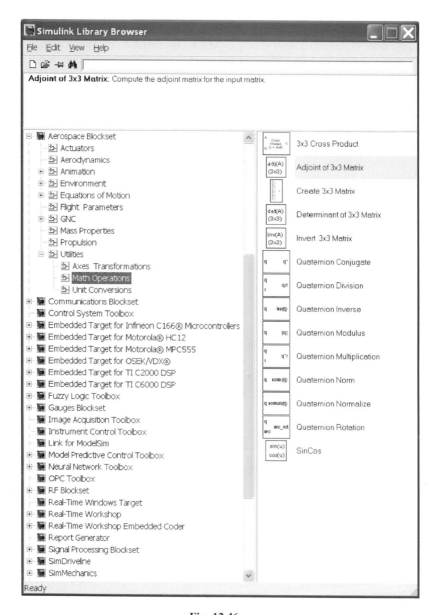

Fig. 12.46

Note that the **Display** block inputs for the adjoint and the inverse are arrays, and so the display is a 3×3 array. To see all of the elements after executing the model, you can resize the block to show more than just the first element. You can resize the block both vertically and horizontally, and the block will add display fields in the appropriate directions. The two black triangles indicate that the block is not displaying all of the input array elements, both in the horizontal and the vertical directions. For example, Fig. 12.47 shows a model that passes a vector (one-dimensional array) to a **Display** block.

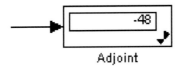

Adjoint

Fig. 12.47

Open the **Invert 3×3 Matrix** block. You will see that it is constructed using **Adjoint of 3×3 Matrix** and **Invert 3×3 Matrix** blocks. It also has an **Assertion** block (see Fig. 12.48).

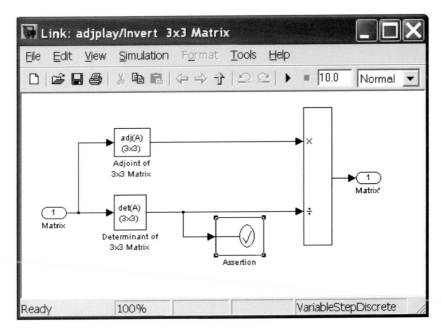

Fig. 12.48

Open **Assertion**, which is used to check for a divide by zero (see Fig. 12.49).

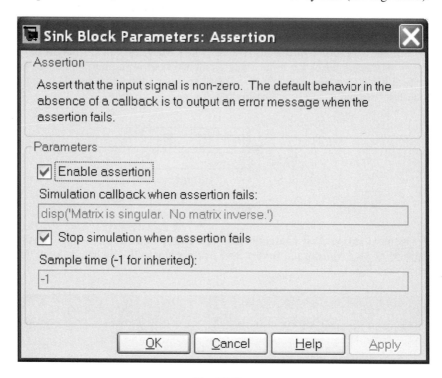

Fig. 12.49

There are two **Configuration Parameters** to set before executing this model (although it will run with warning messages using the defaults). First, select the **discrete Solver** (see Fig. 12.50).

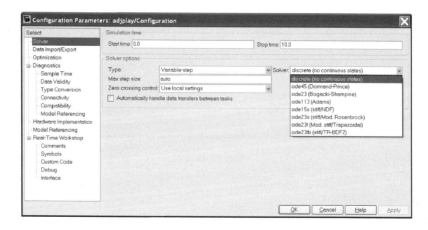

Fig. 12.50

Next, go to the **Diagnostics** menu and change the **Automatic solver parameter selection** diagnostic from **warning** to **none** (see Fig. 12.51).

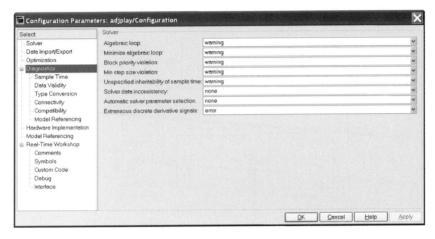

Fig. 12.51

Executing the **adjplay** Simulink model will now give the same results as shown in Secs. 1.8.1 and 1.8.4 without any warning messages.

12.8   Conclusion

This chapter was an introduction to multiple-input, multiple-output systems implemented in Simulink. Also demonstrated was its dynamic modeling capabilities. Some transfer function models were used. Many more elements are available to the user, and much more complex models can be created. Further examples are given in the Simulink manual. Real-time simulations of such systems can be run using the Real-Time Workshop® software.

Other toolboxes are available from The MathWorks to provide additional Simulink modeling elements.

Practice Exercises

12.1   The exercise is to generate and analyze a multi-input, multi-output system. The example is the MIMO representation of a cart and pendulum. The block diagram of the system is shown in Fig. 12.52.

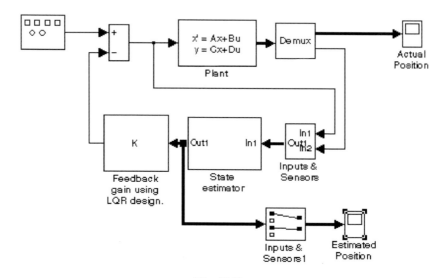

Fig. 12.52

The model parameters that need to be loaded into MATLAB for this model follow:

Plant model constants:

ka = 1
m1 = 1.0
m2 = 1.0

State Estimator Model:
ae =

0	1.0000	−31.4507	0
−1.0000	0	44.8537	0
0	0	−46.6599	1.0000
1.0000	0	−89.5750	0

be =

0	31.4507
1.0000	−43.8537
0	46.6599
0	88.5750

ce =

$$\begin{bmatrix} 1 & 0 & 0 & 0 \\ 0 & 1 & 0 & 0 \\ 0 & 0 & 1 & 0 \\ 0 & 0 & 0 & 1 \end{bmatrix}$$

de =

$$\begin{bmatrix} 0 & 0 \\ 0 & 0 \\ 0 & 0 \\ 0 & 0 \end{bmatrix}$$

Feedback Gain:
K =

$$3.0819 \quad 2.4827 \quad 0.0804 \quad 3.1866$$

The full system representation is shown in Fig. 12.53.

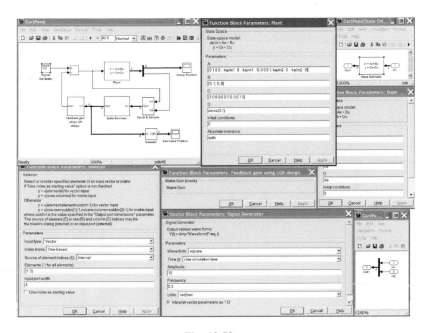

Fig. 12.53

The system response is to a **square** wave with a frequency of 0.3 radians/second and an amplitude of 15. The actual and estimated positions of the response are shown in Fig. 12.54.

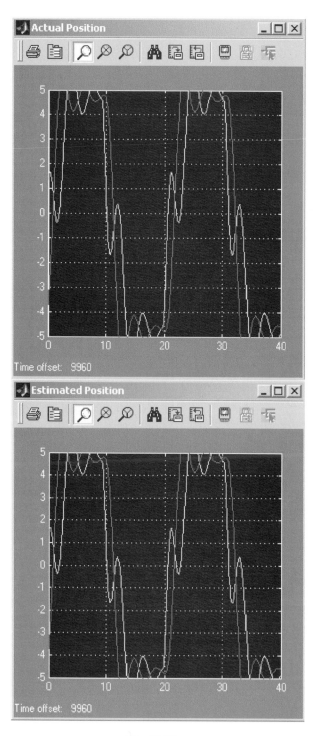

Fig. 12.54

Notes

13
Building Simulink® S-Functions

13.1 Introduction and Objectives

This chapter will introduce the reader to building and using **S-Functions** in Simulink®. Both hand-coded examples, as well as examples generated using the **S-Function Builder**, will be provided.

Upon completion of this chapter, the reader will be able to build Simulink **S-Functions**, use the Simulink **S-Function Builder**, build Simulink models using **S-Functions**, execute and simulate Simulink models including **S-Functions**, and directly interface with these models via MATLAB®.

13.2 Simulink® S-Functions

Using Simulink, you can construct **S-Functions**. An **S-Function** is a computer language description of a Simulink block. **S-Functions** can be written in MATLAB, C, C++, Ada, or FORTRAN. The C, C++, Ada, and FORTRAN **S-Functions** are compiled as MEX-files using the **mex** utility. As with other MEX-files, they are dynamically linked into MATLAB when needed. **S-Functions** use a special calling syntax that enables you to interact with Simulink equation solvers. This interaction is very similar to the interaction that takes place between the solvers and the built-in Simulink blocks.

The form of an **S-Function** is very general and can accommodate continuous, discrete, and hybrid systems. **S-Functions** allow you to add your own blocks to Simulink models. You can create your blocks in MATLAB, C, C++, FORTRAN, or Ada. By following a set of simple rules, you can implement your algorithms in an **S-Function**. After you write your **S-Function** and place its name in an **S-Function** block (see the **User Defined Functions** block library), you can customize the user interface by using masking. You can use **S-Functions** with the Real-Time Workshop® software with constraints. You can also customize the code generated by the Real-Time Workshop software for **S-Functions** by writing a **Target Language Compiler (TLC) TM** file.

335

The **User Defined Functions** block library appears in the **Simulink Library Browser** as shown in Fig. 13.1.

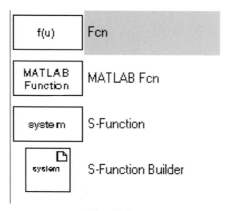

Fig. 13.1

Note that there is a major disadvantage to using **M-file S-Functions**. The MATLAB parser is invoked at every **Simulation** step. This results in a longer simulation run time. Also **M-file S-Functions** cannot be used when code is generated using the Real-Time Workshop software. **C MEX S-Functions** are much faster and can be included in the generated code. A **C MEX S-Function** should be compiled using the **mex** command. This requires a C compiler on the system. Your version of MATLAB may come with the **Lcc C version 2.4** compiler in the directory **Program Files\MATLAB\R2006b\sys\lcc**.

13.3   Simulink® C and S-Function Example, Van der Pol Equation

The following is the model of a second-order nonlinear system. A description of the system can be found on the Simulink diagram in Fig. 13.2. In this section the diagram is shown directly coded into C and then placed into a Simulink **S-Function**.

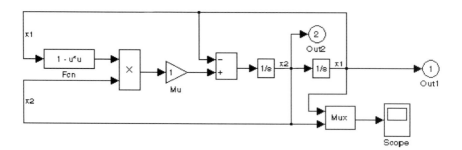

Fig. 13.2

To understand how to implement this system as a Simulink **S-Function**, it is useful to see how an **S-Function** is executed. Execution of a Simulink model proceeds in stages. First comes the initialization phase. In this phase, Simulink incorporates library blocks into the model; propagates widths, data types, and sample times; evaluates block parameters; determines block execution order; and allocates memory. Then Simulink enters a simulation loop. Each pass through the loop is referred to as a simulation step. During each simulation step, Simulink executes each of the model's blocks in the order determined during initialization. For each block, Simulink invokes functions that compute the block's states, derivatives, and outputs for the current sample time. This continues until the simulation is complete. Figure 13.3 illustrates the stages of the execution of a Simulink model containing an **S-Function**.

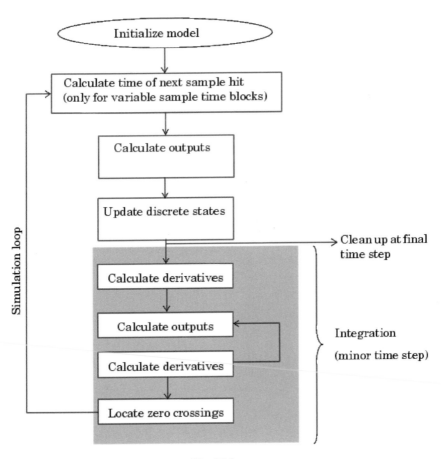

Fig. 13.3

The Van der Pol system implemented as C code appears as follows:

```c
/*  File      : vdpmex.c
 *  Abstract :
 *
 *      Example MEX-file system for Van der Pol equations
 *
 *      Use this as a template for other MEX-file systems
 *      which are only composed of differential equations.
 *
 *      Syntax [sys, x0] = vdpmex(t, x, u, flag)
 *
 *      For more details about S-functions, see simulink/src/sfuntmpl_doc.c
 *
 *  Copyright 1990-2004 The MathWorks, Inc., revised R. Colgren
 *  $Revision: 1.11.4.2.1 $
 */

#define S_FUNCTION_NAME  vdpmex
#define S_FUNCTION_LEVEL 2

#include "simstruc.h"

/*====================*
 * S-function methods *
 *====================*/

/* Function: mdlInitializeSizes
=============================================
 * Abstract:
 *    The sizes information is used by Simulink to determine the S-function
 *    block's characteristics(number of inputs, outputs, states, etc.).
 */
static void mdlInitializeSizes(SimStruct *S)
{
    ssSetNumSFcnParams(S, 0); /* Number of expected parameters */
    if (ssGetNumSFcnParams(S) != ssGetSFcnParamsCount(S)) {
        return; /* Parameter mismatch will be reported by Simulink */
    }

    ssSetNumContStates(S, 2);
    ssSetNumDiscStates(S, 0);

    if (!ssSetNumInputPorts(S, 0)) return;

    if (!ssSetNumOutputPorts(S, 0)) return;

    ssSetNumSampleTimes(S, 1);
    ssSetNumRWork(S, 0);
    ssSetNumIWork(S, 0);
    ssSetNumPWork(S, 0);
    ssSetNumModes(S, 0);
    ssSetNumNonsampledZCs(S, 0);

    /* Take care when specifying exception free code - see sfuntmpl_doc.c */
    ssSetOptions(S, SS_OPTION_EXCEPTION_FREE_CODE);
}
```

```
/* Function: mdlInitializeSampleTimes
==========================================
 * Abstract:
 *    S-function is comprised of only continuous sample time elements
 */
static void mdlInitializeSampleTimes(SimStruct *S)
{
    ssSetSampleTime(S, 0, CONTINUOUS_SAMPLE_TIME);
    ssSetOffsetTime(S, 0, 0.0);
    ssSetModelReferenceSampleTimeDefaultInheritance(S);
}

#define MDL_INITIALIZE_CONDITIONS
/* Function: mdlInitializeConditions
==========================================
 * Abstract:
 *    Initialize both continuous states to zero
 */
static void mdlInitializeConditions(SimStruct *S)
{
    real_T *x0 = ssGetContStates(S);

    /* int x2 */
    x0[0] = 0.25;

    /* int x1 */
    x0[1] = 0.25;
}

/* Function: mdlOutputs
=========================================================
 * Abstract:
 *     This S-Function has no outputs but the S-Function interface requires
 *     that a mdlOutputs() exist so we have a trivial one here.
 */
static void mdlOutputs(SimStruct *S, int_T tid)
{
    UNUSED_ARG(S);   /* unused input argument */
    UNUSED_ARG(tid); /* not used in single tasking mode */
}

#define MDL_DERIVATIVES
/* Function: mdlDerivatives
==================================================
 * Abstract:
 *     xdot(x0) = x0*(1-x1^2) -x1
 *     xdot(x1) = x0
 */
static void mdlDerivatives(SimStruct *S)
{
    real_T *dx = ssGetdX(S);
    real_T *x  = ssGetContStates(S);

    dx[0] = x[0] * (1.0 - x[1] * x[1]) - x[1];
    dx[1] = x[0];
}
```

```
/* Function: mdlTerminate
========================================================
 * Abstract:
 *    No termination needed, but we are required to have this routine.
 */
static void mdlTerminate(SimStruct *S)
{
    UNUSED_ARG(S); /* unused input argument */
}


#ifdef MATLAB_MEX_FILE        /* Is this file being compiled as a MEX-file? */
#include "simulink.c"         /* MEX-file interface mechanism */
#else
#include "cg_sfun.h"          /* Code generation registration function */
#endif
```

Implemented as Simulink **S-Function**, this system appears as shown in Fig. 13.4.

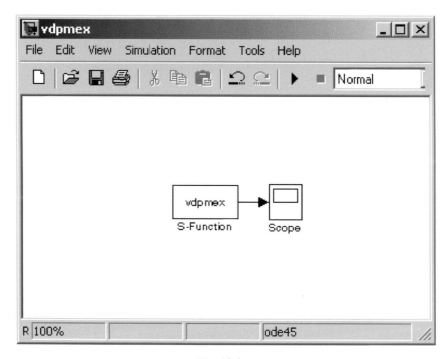

Fig. 13.4

The response of this system as plotted using MATLAB is shown in Fig. 13.5.

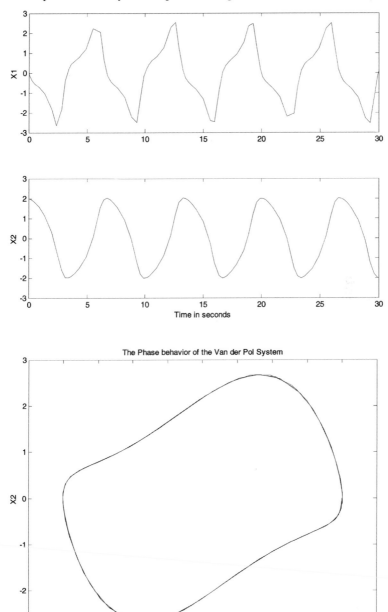

Fig. 13.5

13.4 Simulink® C and S-Function Builder Example, Van der Pol Equation

This section will demonstrate the use of the **Simulink S-Function Builder** to implement the Van der Pol equation of Section 12.3. Before you begin, make sure that you have run **mex–setup** from the MATLAB **Command Window** to choose The MathWorks provided **Lcc C** compiler. Figure 13.6 shows what this looks like on your computer screen.

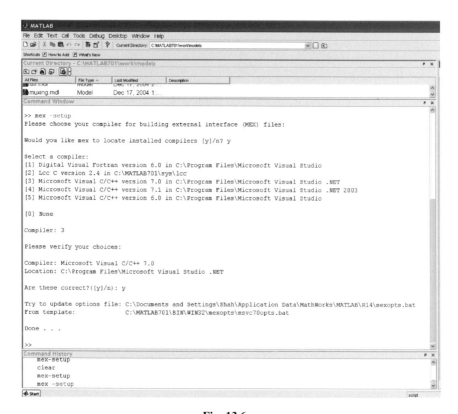

Fig. 13.6

Open up a new Simulink diagram and drag and drop the **S-Function Builder** from the **User-Defined Functions** palette into the new model (see Figs 13.7 and 13.8).

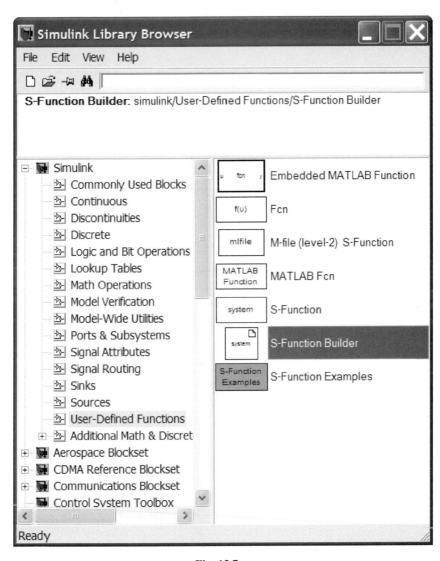

Fig. 13.7

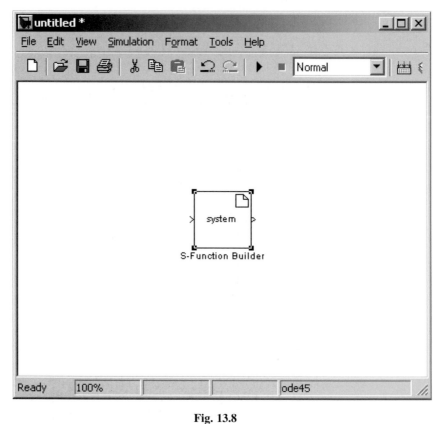

Fig. 13.8

Double-click on the **S-Function Builder** in the model, and you should get the window shown in Fig. 13.9.

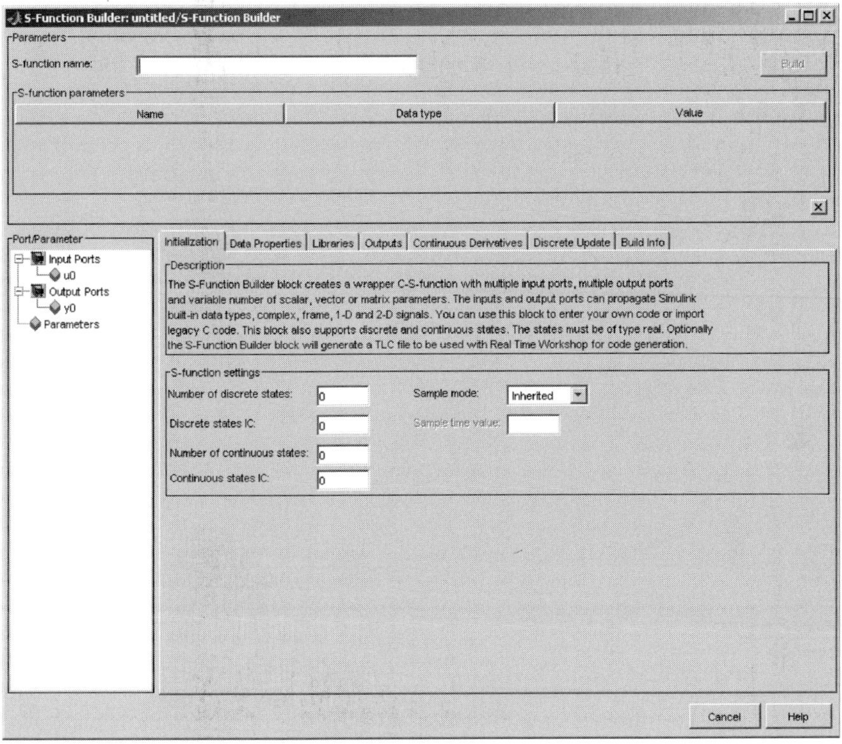

Fig. 13.9

Make the changes shown in Fig. 13.10 to the **Initialization** section.

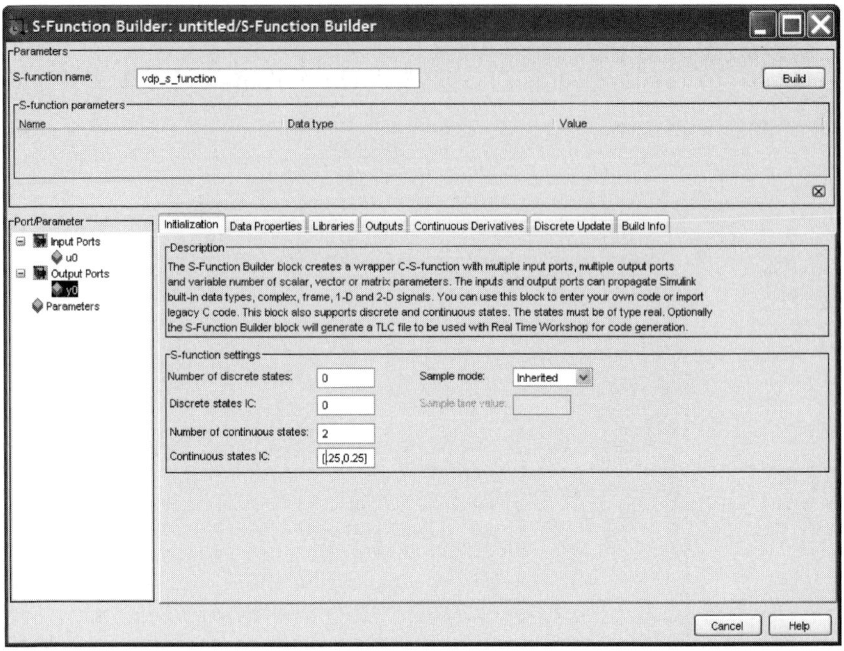

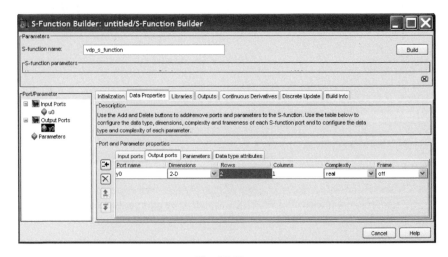

Fig. 13.10

Note that the number of continuous states is **2**, and the continuous states' Initial Conditions (**IC**) are **[0.25,0.25]**. There are no discrete states in this model.

Next select the **Data Properties**=>**Output Ports** tab and change the output to a **2-D** vector and the number of rows to **2**. Pictorially, these selections appear as shown in Fig. 13.11.

Fig. 13.11

There are no changes to the **Libraries** section.
Next select the **Outputs** tab and insert the code shown in Fig. 13.12.

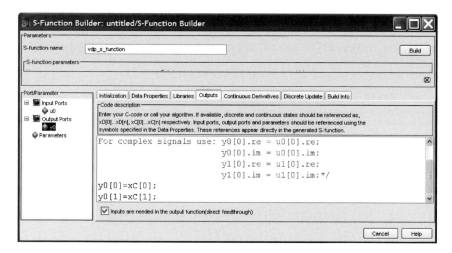

Fig. 13.12

This sets the two outputs to the values of the two continuous states.
Finally, select the **Continuous Derivatives** tab and input the code as shown in Fig. 13.13.

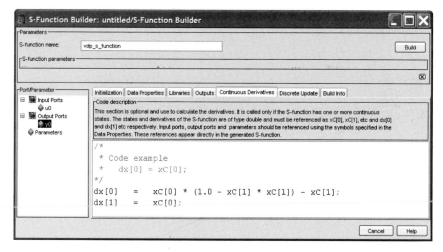

Fig. 13.13

These are the dynamic equations for the Van der Pol equation. These are the same dynamic equations shown in the Simulink diagram at the start of Sec. 13.3.

Next select the **Build Info** tab and then select the **Build** button at the top right corner of the window. You should get a successful build indication. If **mex–setup** was not run from the MATLAB **Command Window** and a C compiler selected, the **Simulink S-Function Builder** would hang up at this point (see Fig. 13.14).

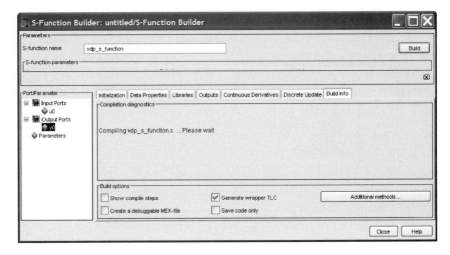

Fig. 13.14

You should now have a **dll** that Simulink can use in the simulation of the model.

The Simulink diagram (after resizing) should now look like that shown in Fig. 13.15.

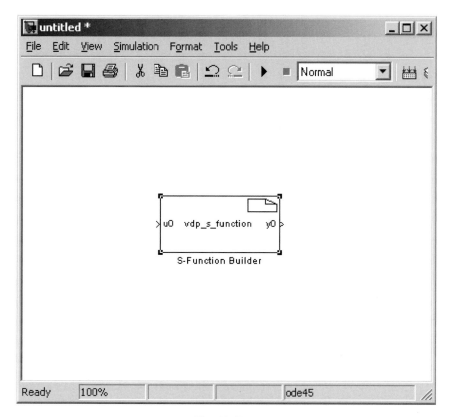

Fig. 13.15

Now select a **Clock** input from the **Sources** palette and a **Scope** from the **Sinks** palette. Also select an **X Y Graph** to duplicate the polar plot generated using MATLAB. To output the data to MATLAB, select a **To Workspace** block. Connect all of these to the output of the **S-Function**.

Save your diagram under the name **sfunvdp.mdl**. Your diagram should now look like the window shown in Fig. 13.16.

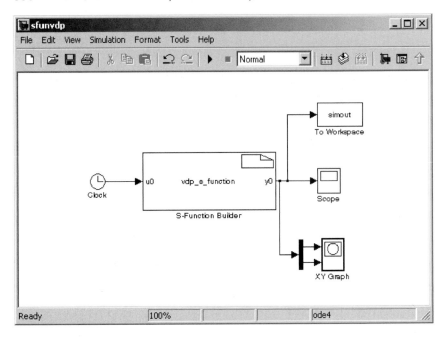

Fig. 13.16

Next you need to choose your simulation parameters. Set the stop time at **20** seconds. Choose a **Fixed-step type solver** using the **Runge-Kutta** integration algorithm (**ode4**) and a **Fixed step size** of **0.01** seconds. After completing this process, select **OK** (see Fig. 13.17).

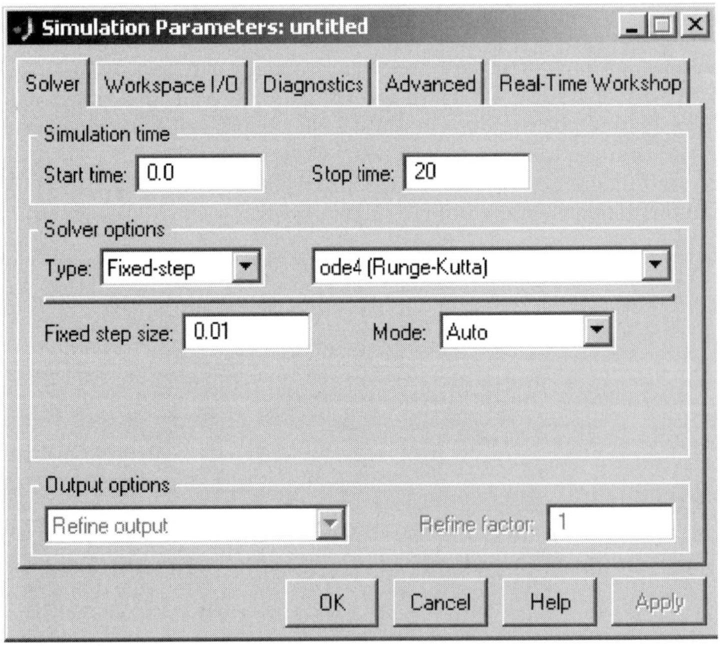

Fig. 13.17

Then run the simulation and check the scope, which should look something like the window shown in Fig. 13.18.

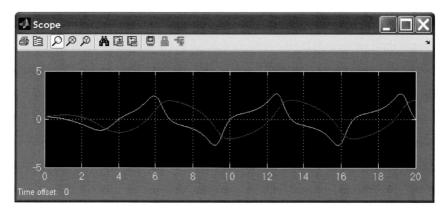

Fig. 13.18

As you sent the **S-Function** output to the workspace using the **To Workspace** block, you can replicate the phase-plane plot using the **plot(simout(:,1), simout(:,2))** command. The plot in Fig. 13.19 shows the response of the system in the phase plane.

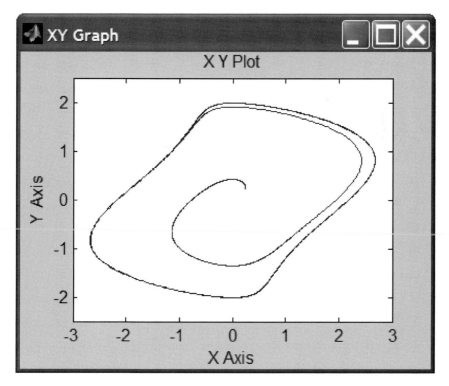

Fig. 13.19

13.5 Example of a FORTRAN S-Function

The following is an example of a FORTRAN **S-Function**. It is a standard model of the Earth's atmosphere. Note that it looks like many of the FORTRAN subroutines that have been developed over the years and are available in libraries for reuse. FORTRAN **S-Functions** allow such code reuse in Simulink.

```
      SUBROUTINE Atmos(alt, sigma, delta, theta)
C
C Calculation of the 1976 standard atmosphere to 86 km.
C This is used to show how to interface Simulink to an
C existing FORTRAN subroutine.
C
C Copyright 1990–2002 The MathWorks, Inc., revised R. Colgren
C
C $Revision: 1.4.1 $
C
      IMPLICIT NONE
C
C --- I/O variables
C
      REAL alt
      REAL sigma
      REAL delta
      REAL theta
C
C --- Local variables
C
      INTEGER i,j,k
      REAL h
      REAL tgrad, tbase
      REAL tlocal
      REAL deltah
      REAL rearth, gmr
C
C --- Initialize values for 1976 atmosphere
C
      DATA rearth/6369.0/! earth radius (km)
      DATA gmr /34.163195/ ! gas constant
C
      REAL htab(8), ttab(8), ptab(8), gtab(8)
      DATA htab/0.0, 11.0, 20.0, 32.0, 47.0, 51.0, 71.0, 84.852/
      DATA ttab/288.15, 216.65, 216.65, 228.65, 270.65, 270.65,
     &      214.65, 186.946/
      DATA ptab/1.0, 2.233611E-1, 5.403295E-2, 8.5666784E-3,
     &       1.0945601E-3, 6.6063531E-4, 3.9046834E-5, 3.68501E-6/
      DATA gtab/-6.5, 0.0, 1.0, 2.8, 0.0, -2.8, -2.0, 0.0/
C
```

```
C --- Convert geometric to geopotential altitude
C
    h = alt*rearth/(alt + rearth)
C
C --- Binary search for altitude interval
    i = 1
    j = 8
C
100 k = (i + j)/2
    IF (h .lt. htab(k)) THEN
     j = k
    ELSE
     i = k
    END IF
    IF ( j .le. i + 1) GOTO 110
    GO TO 100
110 CONTINUE
C
C --- Calculate local temperature
C
    tgrad = gtab(i)
    tbase = ttab(i)
    deltah = h - htab(i)
    tlocal = tbase + tgrad*deltah
    theta = tlocal/ttab(1)
C
C --- Calculate local pressure
C
    IF (tgrad .eq. 0.0) THEN
     delta = ptab(i)*EXP(-gmr*deltah/tbase)
    ELSE
     delta = ptab(i)*(tbase/tlocal)**(gmr/tgrad)
    END IF
C
C --- Calculate local density
C
    sigma = delta/theta
C
    RETURN
    END
```

To incorporate an **S-Function** into a Simulink model, drag an **S-Function** block from the Simulink **User Defined Functions** block library into the model. Then specify the name of the **S-Function** in the **S-Function** name field of the **S-Function** block's dialog box. In this case, the altitude is passed into the subroutine, and the temperature, pressure, and atmospheric density are passed out of the subroutine. There will be one input to and three outputs from this **S-Function** block. This block will appear as shown in Fig. 13.20.

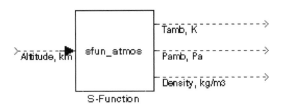

Fig. 13.20

A **CMEX S-Function Gateway** to this FORTRAN routine needs to be written to interface it with the **S-Function** block. This interface is shown in Sec. 13.6.

13.6 Example of a CMEX S-Function Gateway

The following is an example of a Level 2 **CMEX S-Function Gateway**. It allows the FORTRAN standard model of the atmosphere in Sec. 13.5 to be used in an **S-Function**. It is another way that existing FORTRAN code can be made available in libraries for reuse.

```
/*
 *     File: sfun_atmos.c
 *
 *     Abstract: Example of a Level 2 CMEX S-function gateway
 *     to a Fortran subroutine. This technique allows you
 *     to combine the features of level 2 S-functions with
 *     Fortran code, either new or existing.
 *
 *     This example was prepared to be platform neutral.
 *     However, there are portability issues with Fortran
 *     compiler symbol decoration and capitalization (see
 *     prototype section, below).
 *
 *     On Windows using Microsoft Visual C/C++ and Compaq
 *     Visual Fortran 6.0 (a.k.a. Digital Fortran) this
 *     example can be compiled using the following mex
 *     commands (each command is completely on one line):
 *
 *     >> mex -v COMPFLAGS#"$COMPFLAGS /iface:cref" -c
 *        sfun_atmos_sub.f -f ..\..\bin\win32\mexopts\df60opts.bat
 *
 *     >> mex -v
 *        LINKFLAGS#"$LINKFLAGS dformd.lib dfconsol.lib dfport.lib
 *        /LIBPATH:$DF_ROOT\DF98\LIB" sfun_atmos.c sfun_atmos_sub.obj
 *
```

```
*      On linux, one can prepare this example for execution using
*      g77, gcc, and mex:
*
*      % g77 -c sfun_atmos_sub.f -o sfun_atmos_sub.o
*      % mex -lf2c sfun_atmos.c sfun_atmos_sub.o
*
*      or purely with mex on one line:
*
*      >> mex -lf2c sfun_atmos.c sfun_atmos_sub.f
*
*      Gnu Fortran (g77) can be obtained for free from many
*      download sites, including http://www.redhat.com in
*      the download area. Keyword on search engines is 'g77'.
*
*      R. Aberg, 01 JUL 2000, revised R. Colgren
*      Copyright 1990-2005 The MathWorks, Inc.
*
*      $Revision: 1.8.4.5.1 $
*/
#define S_FUNCTION_NAME sfun_atmos
#define S_FUNCTION_LEVEL 2

#include "simstruc.h"

/*
 * Below is the function prototype for the Fortran
 * subroutine 'Atmos' in the file sfun_atmos_sub.f.
 *
 * Note that datatype REAL is 32 bits in Fortran,
 * so the prototype arguments must be float.
 *
 * Your Fortran compiler may decorate and/or change
 * the capitalization of 'SUBROUTINE Atmosphere'
 * differently than the prototype below. Check
 * your Fortran compiler's manual for options to
 * learn about and possibly control external symbol
 * decoration.
 *
 * Additionally, you may want to use CFortran,
 * a tool for automating the interface generation
 * between C and Fortran ... in either direction.
 * Search the web for 'cfortran'.
 */
/*
 * Digital Fortran's external symbols are in capitals
 * on Windows platforms; preceding underscore is implicit.
 */

#if defined(_WIN32) && ! defined(_WIN64)
#define atmos_ ATMOS
#endif

/*
 * Note that some compilers don't use a trailing
 * underscore on Fortran external symbols
 */
```

```c
#if defined(__xlc__) || defined(__hpux) || defined(_WIN64)
#define atmos_ atmos
#endif

extern void atmos_(float *alt,
                   float *sigma,
                   float *delta,
                   float *theta);
/* Parameters for this block */

typedef enum {T0_IDX=0, P0_IDX, R0_IDX, NUM_SPARAMS } paramIndices;

#define T0(S) (ssGetSFcnParam(S, T0_IDX))
#define P0(S) (ssGetSFcnParam(S, P0_IDX))
#define R0(S) (ssGetSFcnParam(S, R0_IDX))

/* Function: mdlInitializeSizes
========================================================
 * Abstract:
 *     Set up the sizes of the S-function's
 *     inputs and outputs.
 */

static void mdlInitializeSizes(SimStruct *S)
{
    ssSetNumSFcnParams(S,NUM_SPARAMS); /* expected number */
#if defined(MATLAB_MEX_FILE)
    if (ssGetNumSFcnParams(S) != ssGetSFcnParamsCount(S)) goto EXIT_POINT;
#endif
    {
        int iParam = 0;
        int nParam = ssGetNumSFcnParams(S);

        for ( iParam = 0; iParam < nParam; iParam++ )
        {
            ssSetSFcnParamTunable( S, iParam, SS_PRM_SIM_ONLY_TUNABLE );
        }
    }

    ssSetNumContStates( S, 0 );
    ssSetNumDiscStates( S, 0 );

    ssSetNumInputPorts(S, 1);
    ssSetInputPortWidth(S, 0,   DYNAMICALLY_SIZED);
    ssSetInputPortDirectFeedThrough(S, 0, 1);
    ssSetInputPortRequiredContiguous(S, 0, 1);

    ssSetNumOutputPorts(S, 3);
    ssSetOutputPortWidth(S, 0, DYNAMICALLY_SIZED); /* temperature */
    ssSetOutputPortWidth(S, 1, DYNAMICALLY_SIZED); /* pressure    */
    ssSetOutputPortWidth(S, 2, DYNAMICALLY_SIZED); /* density     */

EXIT_POINT:
    return;
}
```

```
/*   Function:   mdlInitializeSampleTimes
========================================================
 * Abstract:
 *    Specify that we inherit our sample time from
 *    the driving block.
 */
static void mdlInitializeSampleTimes(SimStruct *S)
{
    ssSetSampleTime(S, 0, INHERITED_SAMPLE_TIME);
    ssSetOffsetTime(S, 0, 0.0);
    ssSetModelReferenceSampleTimeDefaultInheritance(S);
}
/*   Function: mdlOutputs
========================================================
 * Abstract:
 *    Calculate atmospheric conditions using Fortran subroutine.
 */
static void mdlOutputs(SimStruct *S, int_T tid)
{
    double *alt = (double *) ssGetInputPortSignal(S,0);
    double *T   = (double *) ssGetOutputPortRealSignal(S,0);
    double *P   = (double *) ssGetOutputPortRealSignal(S,1);
    double *rho = (double *) ssGetOutputPortRealSignal(S,2);
    int    w  = ssGetInputPortWidth(S,0);
    int    k;
    float falt, fsigma, fdelta, ftheta;

    for (k=0; k<w; k++) {

      /* set the input value */
      falt = (float) alt[k];

      /* call the Fortran routine using pass-by-reference */
      atmos_(&falt, &fsigma, &fdelta, &ftheta);

      /* format the outputs using the reference parameters */
      T[k] = mxGetScalar(T0(S)) * (double) ftheta;
      P[k] = mxGetScalar(P0(S)) * (double) fdelta;
      rho[k] = mxGetScalar(R0(S)) * (double) fsigma;
    }
}


/*   Function:   mdlTerminate
========================================================
 * Abstract:
 *    This method is required for Level 2 S-functions.
 */
static void mdlTerminate(SimStruct *S)
{
}

#ifdef MATLAB_MEX_FILE    /* Is this file being compiled as a MEX-file? */
#include "simulink.c"     /* MEX-file interface mechanism */
#else
#include "cg_sfun.h"      /* Code generation registration function */
#endif
```

To incorporate an **S-Function** into a Simulink model, drag an **S-Function** block from the Simulink **User Defined Functions** block library into the model. Then specify the name of the **S-Function** in the **S-Function** name field of the **S-Function** block's dialog box, as illustrated in Fig. 13.21.

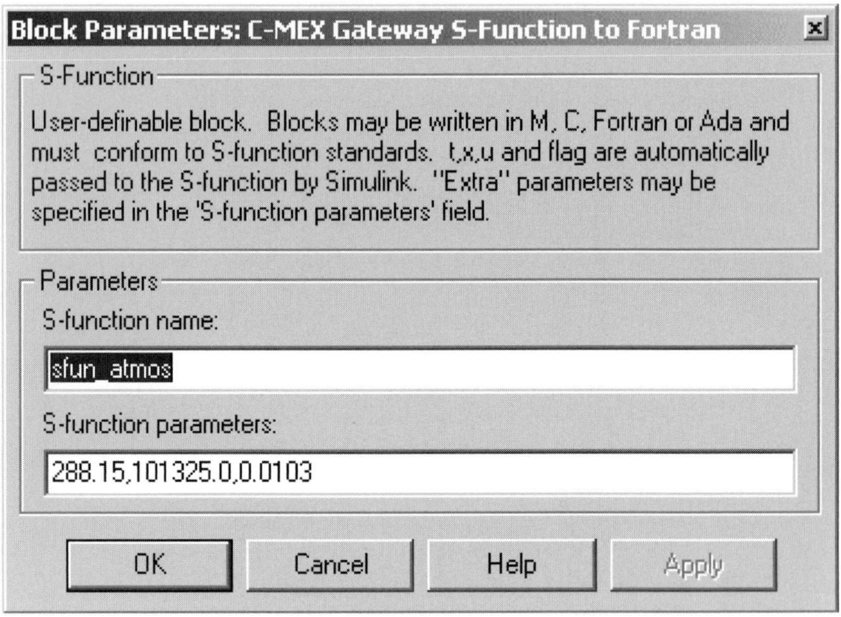

Block Parameters: C-MEX Gateway S-Function to Fortran   ⊠

┌─ S-Function ───┐
│ User-definable block. Blocks may be written in M, C, Fortran or Ada and │
│ must conform to S-function standards. t,x,u and flag are automatically │
│ passed to the S-function by Simulink. "Extra" parameters may be │
│ specified in the 'S-function parameters' field. │
└───┘

┌─ Parameters ───┐
│ S-function name: │
│ ┌───┐ │
│ │ sfun_atmos │ │
│ └───┘ │
│ S-function parameters: │
│ ┌───┐ │
│ │ 288.15,101325.0,0.0103 │ │
│ └───┘ │
└───┘

| OK | Cancel | Help | Apply |

Fig. 13.21

Note that extra parameters may be specified in the **S-Function** parameter field.

You can also use the **S-Function** builder located in the **Simulink Library Browser** to help you build a Simulink **S-Function**.

13.7 Simulink® Block Diagram Using S-Function

The following shows a Simulink block diagram constructed using this **S-Function**. For this example a single scope with four outputs was constructed. It is executed in the same way as any other Simulink model. The FORTRAN code and the **CMEX S-Function Gateway** are both embedded as text into the diagram for reference. The color of these text blocks are modified from that of the rest of the diagram using **Format, Background Color, Light Blue** for highlighting. The result of this simulation is shown in Figs 13.22 and 13.23.

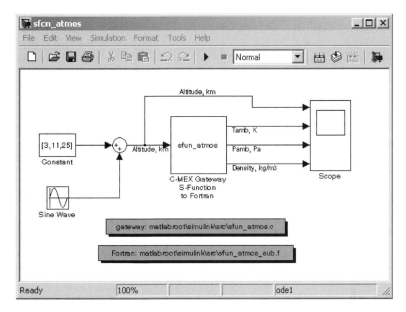

Fig. 13.22

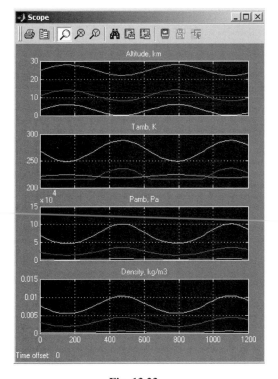

Fig. 13.23

13.8 Conclusion

This chapter was an introduction to building and using **S-Functions** in Simulink. Examples were provided where the entire **S-Function** was hand coded or where it was generated using the **S-Function Builder**. A **CMEX S-Function Gateway** for a FORTRAN **S-Function** was also shown. Further examples are given in the **Simulink Library Browser**.

Practice Exercises

13.1 The **S-Function** to be generated in this exercise is a two-input, two-output continuous-time model programmed in C. A hand-programmed copy of this model is found in the **Simulink Library Browser** under **S-Function demos** in the **C-file S-Functions** under **Continuous**. Select the model **Continuous time system**. Use the system model **Continuous-time state space S-Function**. Excite the system using two signal generators. You will have two outputs. You will need a **Mux** bus to make a two-dimensional signal vector to drive the system. Look at the system output using a **Scope**.

From the following code, use the **S-Function Generator** to generate this **S-Function** as shown in this chapter.

To test this **S-Function**, start with two **sine** wave inputs at a frequency of 1 rad/second and an amplitude of 1. Then try other excitation signals and examine the response (see Fig. 13.24).

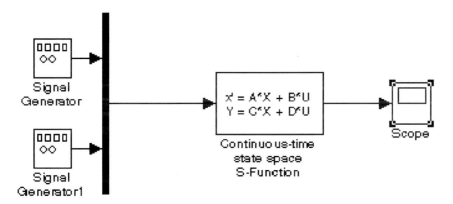

Fig. 13.24

The C code for this system follows:

```
/*   File : csfunc.c
 *   Abstract: Example C-file S-function for defining a continuous system.
 *
 *       xdot = Ax + Bu
 *       y    = Cx + Du
 *
 *   For more details about S-functions, see simulink/src/sfuntmpl_doc.c.
 *   Copyright 1990-2004 The MathWorks, Inc., revised R. Colgren
 *   $Revision: 1.9.4.2.1 $
 */


#define S_FUNCTION_NAME csfunc
#define S_FUNCTION_LEVEL 2


#include "simstruc.h"


#define U(element) (*uPtrs[element]) /* Pointer to Input Port0 */


static real_T A[2][2]={ { -0.09, -0.01 } ,
                        {  1   ,  0    }
                      };


static real_T B[2][2]={ {  1   , -7    } ,
                        {  0   , -2    }
                      };


static real_T C[2][2]={ {  0   ,  2    } ,
                        {  1   , -5    }
                      };


static real_T D[2][2]={ { -3   ,  0    } ,
                        {  1   ,  0    }
                      };
/*===================*
 * S-function methods *
 *===================*/


/* Function: mdlInitializeSizes
=========================================================
 * Abstract:
 *    The sizes information is used by Simulink to determine the S-function
 *    block's characteristics (number of inputs, outputs, states, etc.).
 */
static void mdlInitializeSizes(SimStruct *S)
{
    ssSetNumSFcnParams(S, 0); /* Number of expected parameters */
    if (ssGetNumSFcnParams(S) ! = ssGetSFcnParamsCount(S)) {
        return; /* Parameter mismatch will be reported by Simulink */
    }
```

```
    ssSetNumContStates(S, 2);
    ssSetNumDiscStates(S, 0);

    if (!ssSetNumInputPorts(S, 1)) return;
    ssSetInputPortWidth(S, 0, 2);
    ssSetInputPortDirectFeedThrough(S, 0, 1);

    if (!ssSetNumOutputPorts(S, 1)) return;
    ssSetOutputPortWidth(S, 0, 2);

    ssSetNumSampleTimes(S, 1);
    ssSetNumRWork(S, 0);
    ssSetNumIWork(S, 0);
    ssSetNumPWork(S, 0);
    ssSetNumModes(S, 0);
    ssSetNumNonsampledZCs(S, 0);

    /* Take care when specifying exception free code - see sfuntmpl_doc.c */
    ssSetOptions(S, SS_OPTION_EXCEPTION_FREE_CODE);
}

/* Function: mdlInitializeSampleTimes
 ======================================================
 * Abstract:
 *    Specifiy that we have a continuous sample time.
 */
static void mdlInitializeSampleTimes(SimStruct *S)
{
    ssSetSampleTime(S, 0, CONTINUOUS_SAMPLE_TIME);
    ssSetOffsetTime(S, 0, 0.0);
    ssSetModelReferenceSampleTimeDefaultInheritance(S);
}


#define MDL_INITIALIZE_CONDITIONS
/* Function: mdlInitializeConditions
 ======================================================
 * Abstract:
 *    Initialize both continuous states to zero.
 */
static void mdlInitializeConditions(SimStruct *S)
{
   real_T *x0 = ssGetContStates(S);
   int_T lp;

   for (lp=0;lp<2;lp++) {
       *x0++=0.0;
   }
}

/* Function: mdlOutputs
 ======================================================
 * Abstract:
 *        y = Cx + Du
 */
```

```
static void mdlOutputs(SimStruct *S, int_T tid)
{
    real_T         *y        = ssGetOutputPortRealSignal(S,0);
    real_T         *x        = ssGetContStates(S);
    InputRealPtrsType uPtrs = ssGetInputPortRealSignalPtrs(S,0);

    UNUSED_ARG(tid); /* not used in single tasking mode */

    /* y = Cx + Du */
    y[0]=C[0][0]*x[0]+C[0][1]*x[1]+D[0][0]*U(0)+D[0][1]*U(1);
    y[1]=C[1][0]*x[0]+C[1][1]*x[1]+D[1][0]*U(0)+D[1][1]*U(1);
}


#define MDL_DERIVATIVES
/* Function: mdlDerivatives
==========================================================
 * Abstract:
 *         xdot = Ax + Bu
 */
static void mdlDerivatives(SimStruct *S)
{
    real_T         *dx       = ssGetdX(S);
    real_T         *x        = ssGetContStates(S);
    InputRealPtrsType uPtrs = ssGetInputPortRealSignalPtrs(S,0);

    /* xdot=Ax+Bu */
    dx[0]=A[0][0]*x[0]+A[0][1]*x[1]+B[0][0]*U(0)+B[0][1]*U(1);
    dx[1]=A[1][0]*x[0]+A[1][1]*x[1]+B[1][0]*U(0)+B[1][1]*U(1);
}


/* Function: mdlTerminate
==========================================================
 * Abstract:
 *    No termination needed, but we are required to have this routine.
 */
static void mdlTerminate(SimStruct *S)
{
    UNUSED_ARG(S); /* unused input argument */
}


#ifdef MATLAB_MEX_FILE    /* Is this file being compiled as a MEX-file? */
#include "simulink.c"      /* MEX-file interface mechanism */
#else
#include "cg_sfun.h"       /* Code generation registration function */
#endif
```

Notes

Basic Stateflow®

14

Introduction to Stateflow®

14.1   Introduction and Objectives

This final chapter introduces the reader to The MathWorks' Stateflow®
graphical modeling capabilities. It assumes a basic familiarity with
MATLAB® and Simulink®.

Upon completion of this chapter, the reader will be able to identify the
graphics modeling capabilities of Stateflow software, open and close Stateflow
models from MATLAB, generate Stateflow models, execute and simulate
systems implemented within Stateflow software, and modify Stateflow model
block parameters from Simulink and MATLAB.

14.2   Opening, Executing, and Saving Stateflow® Models

Stateflow software is an excellent tool to dynamically simulate switching and
other state changes within MATLAB's graphical tools, including Simulink.
Although Stateflow software was originally developed as a stand-alone graphical
modeling environment, it is now combined within Simulink. This offers the
important advantage that Simulink blocks can be used with Stateflow blocks
within the same model.

To provide the reader with an example of the appearance of a Stateflow model,
we will open the following: 1) a Simulink model containing a Stateflow block and
2) the Stateflow library.

First we will open a Simulink model from the MATLAB **Command Window**.
The model we will open is a simple demonstration of **While** and **Do While** loops
modeled within Stateflow software. To open this model titled **sf_while.mdl**,
simply type the following within the MATLAB **Command Window**:

> » **sf_while**

The Simulink model window shown in Fig. 14.1 is opened.

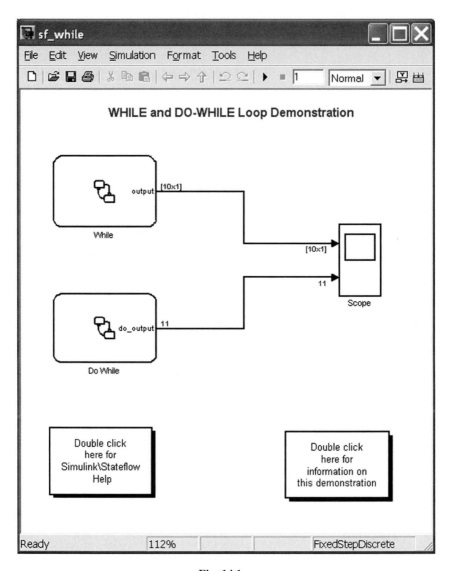

Fig. 14.1

Note that this model is set to run using a fixed-step discrete integration routine. To view the Stateflow model called **While**, simply double-click on this block. The Stateflow model shown in Fig. 14.2 is opened.

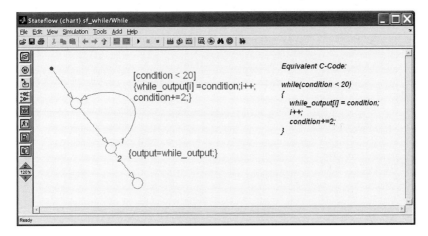

Fig. 14.2

Similarly, to open the Stateflow model called **Do While**, double-click on the block with this name. This will open the Stateflow model in Fig. 14.3.

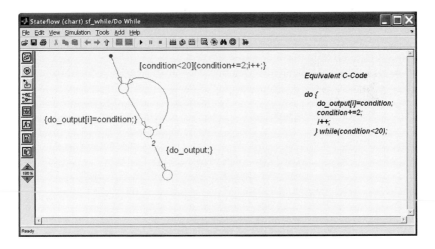

Fig. 14.3

You will next execute this model from Simulink in the same way that you previously ran Simulink models. For example, you can simply click on the **Start Simulation** symbol at the center of the Simulink toolbar. When you run the

Simulink model **sf_while**, you will note that the active state flow paths will show a widened font, whereas the inactive state flow paths will show the standard line width. This is demonstrated in Figs 14.4 and 14.5. Figure 14.4 shows the initial entry into the **While** Stateflow chart, before it does the conditional test.

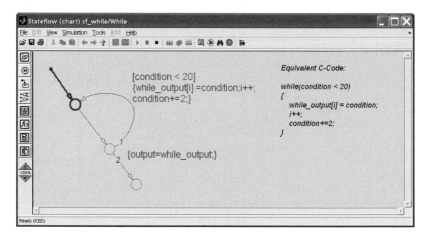

Fig. 14.4

The second Stateflow chart (Fig. 14.5) shows the **While** conditional test being invoked.

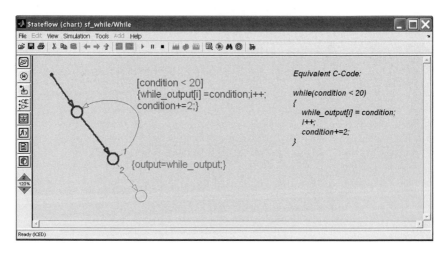

Fig. 14.5

The case where the **output=while_output** is shown in Fig. 14.6.

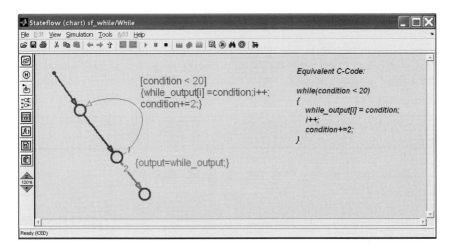

Fig. 14.6

Similarly, the case where **do_output** is invoked is shown in Fig. 14.7.

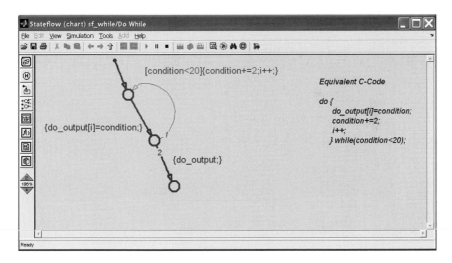

Fig. 14.7

Note that the simulation's end time is set to only 1 second for this demonstration. To provide sufficient time to fully observe this model executing, let us reset the simulation run time to end at 100 seconds by restarting the simulation from the MATLAB **Command Window**. This is accomplished using the following command:

>> **sim('sf_while',100);**

If you open the **Scope**, you will see the time offset advance to 100 seconds in the lower left-hand corner of the **Scope** plot window. In Fig. 14.8, the simulation has almost finished executing and displays an execution time of 96 seconds.

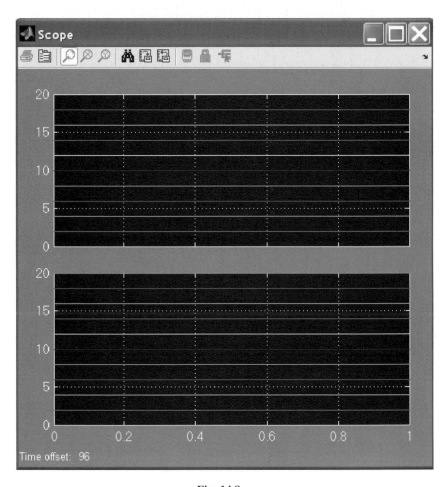

Fig. 14.8

The simulation will stop before executing at a time of 100 seconds, and so the plot window will appear cleared of the vector of values at the completion of the simulation run. This result is shown in Fig. 14.9.

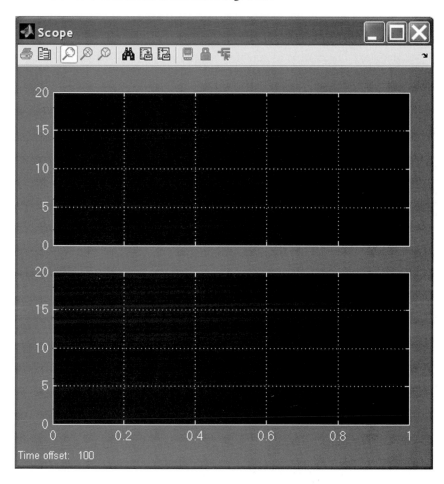

Fig. 14.9

To save this model within your MATLAB working directory, simply invoke the **save_system** command from the MATLAB **Command Window** as follows:

> ≫ **save_system('sf_while');**

To close this model, simply invoke the **close_system** command from the MATLAB **Command Window** as follows:

> ≫ **close_system('sf_while');**

14.3   Constructing a Simple Stateflow® Model

A Stateflow model is a version of a finite state machine for controlling a physical plant. A finite state machine is a representation of an event-driven system. In an event-driven system, the system responds by making a transition from one state to another state in response to an event. This occurs when the condition defining the change is set to true from false.

A Stateflow diagram is a graphical representation of such a finite state machine, where states and transitions form the basic building blocks of the system. You can also represent signal flows as stateless diagrams using Stateflow software. Stateflow software provides you with the elements and construction tools that you need to include states and transitions within a Simulink model. Starting with Stateflow Version 14 Service Pack 3, you are also provided with a **Truth Table** block.

Stateflow charts are often used to control a physical plant in response to events such as a temperature or pressure change. The physical plant can also be controlled based on user-driven events. For example, you can use a state machine to represent the gear selection process in a car's automatic transmission. The transmission has a number of operating states: park, reverse, neutral, drive, and low. As the driver shifts from one position to another, the system makes a transition from one state to another. Examples of such user-driven events are shifting from park to reverse, from reverse to drive, from drive to neutral, etc.

The first step in generating a Stateflow model is to invoke the Stateflow program, which opens the Stateflow library window. This is accomplished by typing

>> **stateflow**

from within the MATLAB **Command Window**.

The Stateflow block library **sflib** appears as in Fig. 14.10.

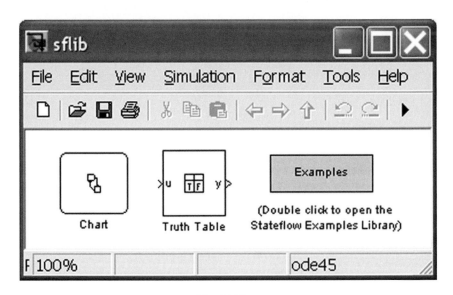

Fig. 14.10

A Stateflow diagram is created using the block on the left-hand side of **sflib**. A **Truth Table** generator is provided in the center of this window. Stateflow examples are included within the library block on the right-hand side of this window.

To start, let us create a new Simulink model and drag a copy of the Stateflow **Chart** block into this model's window. This can be done from **sflib** by going to **File**, **New**, **Model** as in Fig. 14.11.

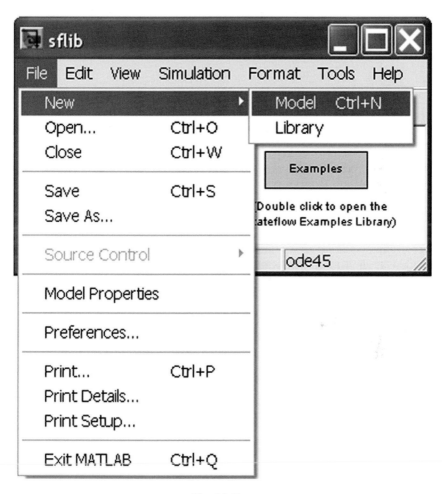

Fig. 14.11

An untitled Simulink model window appears as in Fig. 14.12.

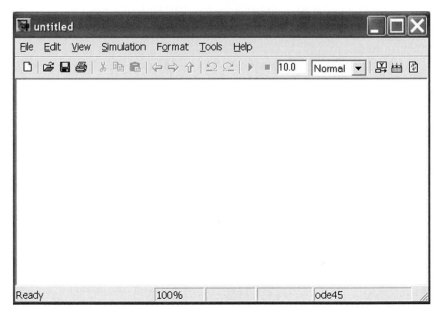

Fig. 14.12

Next drag a copy of the Stateflow block **Chart** onto the **untitled** model window as shown in Fig. 14.13. Note the asterisk ***** after **untitled** in the **Title bar**, denoting an unsaved Simulink model.

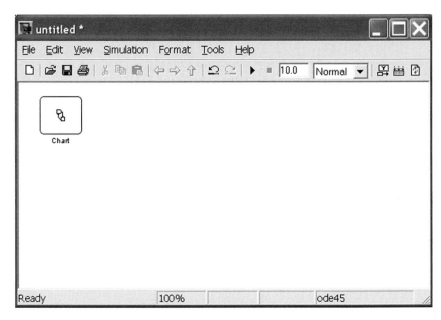

Fig. 14.13

Next double-click on the Stateflow block **Chart** within the **untitled** Simulink model to invoke a **Stateflow diagram editor** window. The Stateflow editor then appears as in Fig. 14.14.

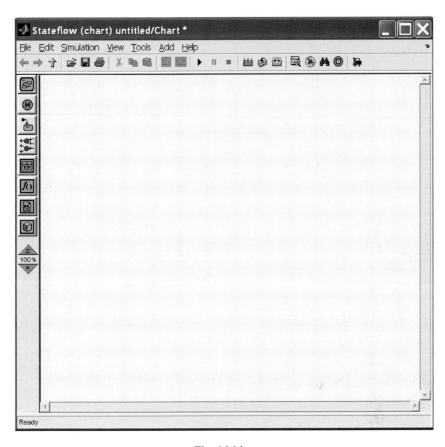

Fig. 14.14

Next we will construct a simple on-off switch using Stateflow software and will use a simple **Manual Switch** from Simulink to manually toggle the states. The on-off switch will look like that in Fig. 14.15. The Simulink diagram driving this switch is also shown.

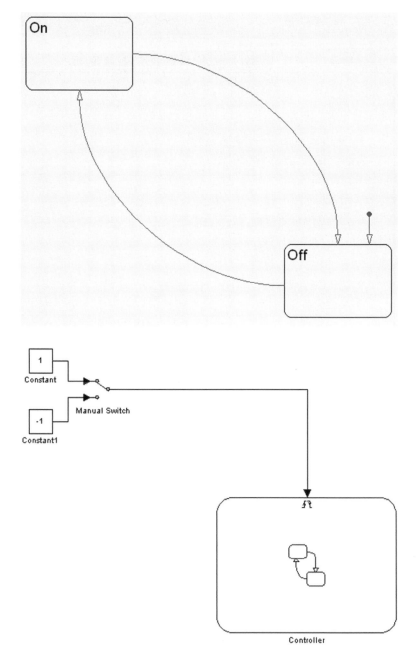

Fig. 14.15

First we need to insert two states into our model. One will represent the switch being on; the other will represent the switch being off. States are accessed using the upper blue **State** tool icon 🔲 in the **drawing toolbar** or **object palette** on the left side of the Stateflow model window. The **object palette** contains a set of tools for drawing **states**, a **history junction**, a **default transition**, a **connective junction**, a **truth table**, a **function**, an **embedded MATLAB function**, and a **box** (see Fig. 14.16).

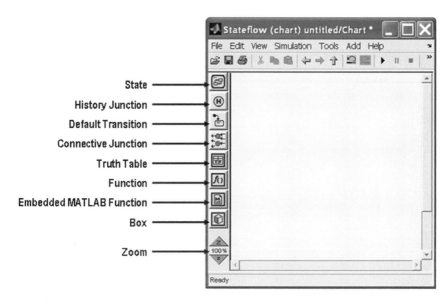

State
History Junction
Default Transition
Connective Junction
Truth Table
Function
Embedded MATLAB Function
Box
Zoom

Fig. 14.16

Next the state transitions need to be added. Transitions originate with a source state and terminate at a destination state. Evoking a transition means that the source state has become inactive and the destination state has become active. In our switch model, the **On** state becomes active by moving through the transition from the **Off** state to the **On** state. Once on, the switch goes to off by moving through an **On** to **Off** transition. In our case, we will start our model with the switch in the **Off** state. This is accomplished by using a default transition (see Fig. 14.17).

Fig. 14.17

Now that we have moved a **State** into the Stateflow **Chart**, we need to label this **State**. We will make this **State** our **On** state and will label it accordingly. Note that the **State** block changes to a pink color while it is being edited. It reverts to black after it is properly labeled. Later you will see that when the simulation is being executed, the active **State** block is shown in bold blue. These default colors can be modified as desired by the user (see Fig. 14.18).

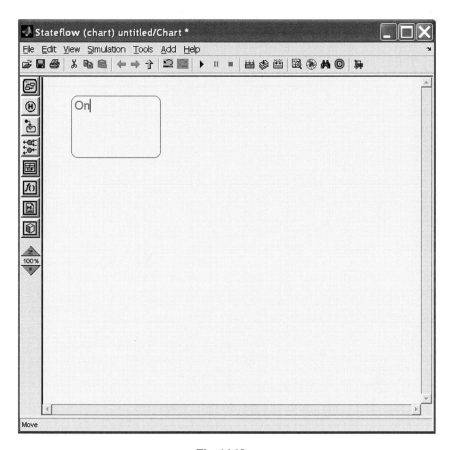

Fig. 14.18

The same procedure is used to create the **Off** state in our model. Again, while the **State** is being modified, it is shown in pink. Note that the previously created **On** state is shown in black. As the Simulink model **untitled** has not yet been saved, the **Title bar** shows an asterisk ***** after the name **untitled/Chart** given as a default to this Stateflow model (see Fig. 14.19).

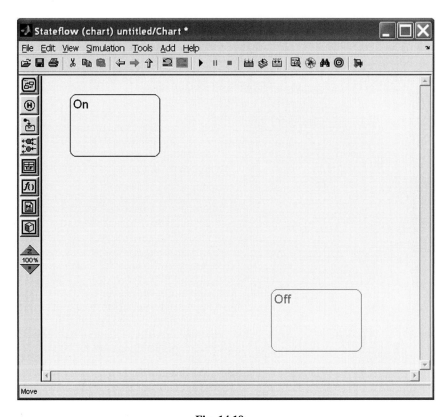

Fig. 14.19

Next we will draw our first state transition, from the **Off** state to the **On** state. The **Off** state and the **state transition** are both shown in pink because they are in the process of being modified. The **state transition** is simply drawn by clicking on the starting **State** and holding down the left mouse button until the border of the concluding **State** is reached. To start the **SMART mode**, just hold down the **s** key while dragging the **state transition** (see Fig. 14.20).

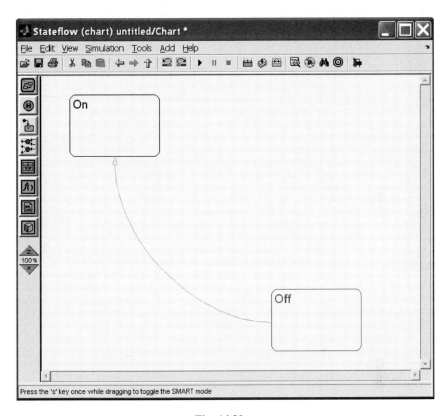

Fig. 14.20

Our next task is to draw the state transition from the **On** state to the **Off** state. The **On** state and the state transition are now both shown in pink, again because they are in the process of being modified (see Fig. 14.21).

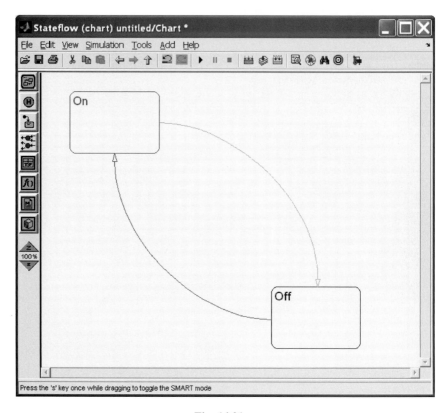

Fig. 14.21

Our final transition is the **default transition**. We will set the **State Off** to be the default state. Select the **default transition** (third item from the top in the **object palette**) and drag it to the top of the **Off** state block. As it is being modified, it will change from blue (Fig. 14.22) to pink (Fig. 14.23). To complete the **default transition**, we will need to connect it with the higher-level Simulink diagram **untitled**.

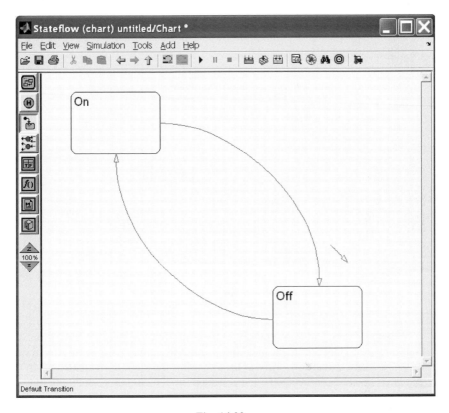

Fig. 14.22

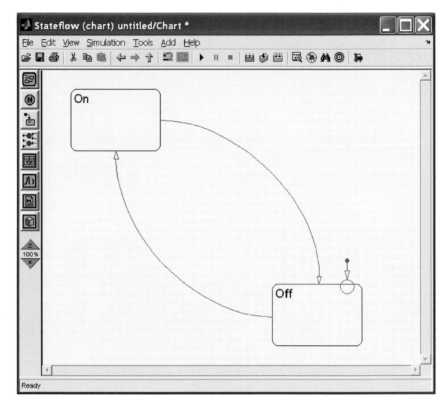

Fig. 14.23

When released, and thus attached to the **State Off**, the **default transition** will continue to be shown in pink. In addition, a question mark will be displayed next to the **default transition**. This is to show that the **default transition** has not yet been properly connected to the trigger in the Simulink diagram untitled (see Fig. 14.24).

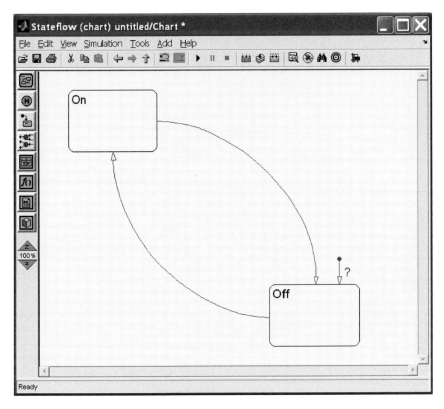

Fig. 14.24

We will next use a simple manual switch from Simulink to manually toggle the states. Locate the **Chart** block in the lower right corner of the untitled diagram to make room for the **Manual Switch**. This action is shown in Fig. 14.25.

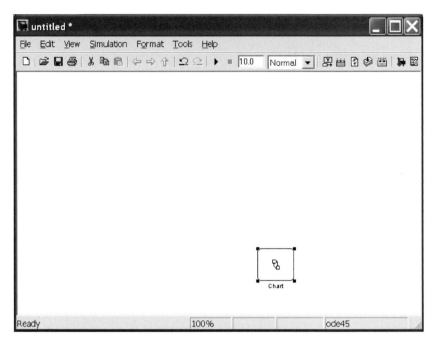

Fig. 14.25

Enlarge the **Chart** block to make room for the **Trigger** symbol as shown in the diagram at the start of this section (see Fig. 14.26).

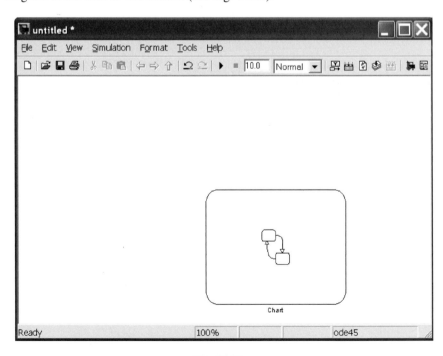

Fig. 14.26

Next name the **Chart** block **Controller** as in Fig. 14.27.

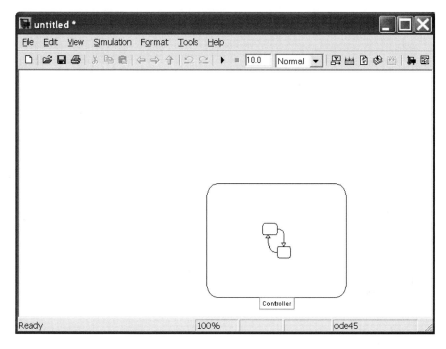

Fig. 14.27

Two **Constant** blocks need to be added to the untitled window. Select a **Constant** from the Simulink Library Browser as shown in Fig. 14.28.

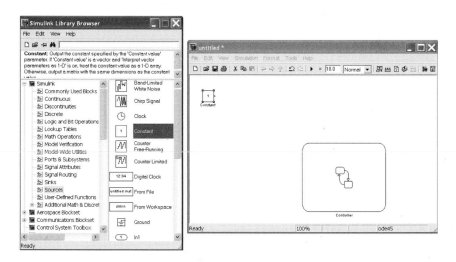

Fig. 14.28

The two **Constant** blocks are next shown after being located in the upper left corner of the **untitled** diagram. As the first **Constant** block was left with the default name, Simulink automatically renames the second block to be **Constant1** (see Fig. 14.29).

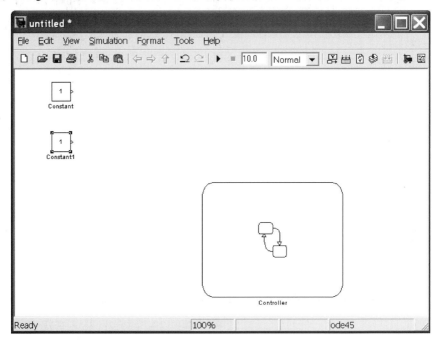

Fig. 14.29

The final Simulink block required is a **Manual Switch**. Locate a **Manual Switch** under the Signal Routing Simulink Library and drag a copy into the **untitled** model as shown in Fig. 14.30.

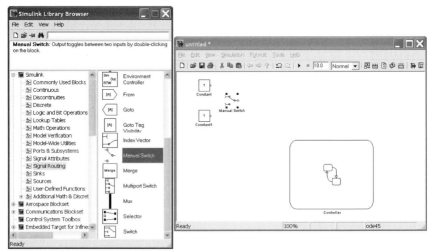

Fig. 14.30

Next start the process of connecting the Simulink blocks (see Fig. 14.31).

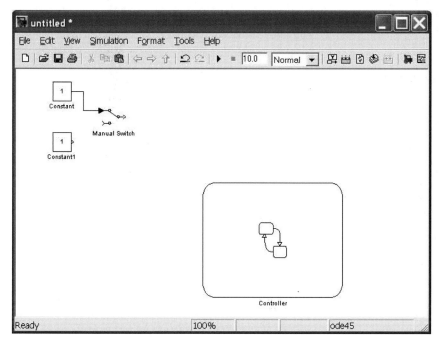

Fig. 14.31

The two **Constant** blocks are connected to the two sides of the **Manual Switch** as we have done previously. The first connection method involves holding down the **Ctrl** key while first clicking on the originating block and then clicking on the completing box. The second method requires that the first block be selected by clicking on the first block; then the path is completed by dragging the signal path to the desired input location (see Fig. 14.32).

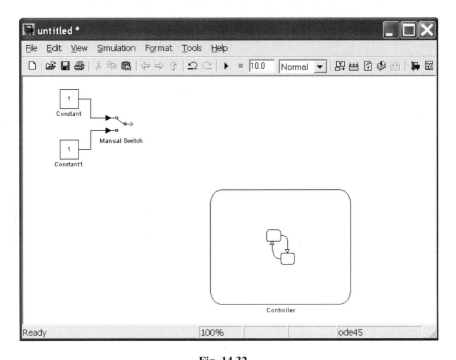

Fig. 14.32

Next we need to go into the **Controller** block and add an input from Simulink to set the **State** blocks based on the **Manual Switch**. To accomplish this, first select the **Add** pull-down menu, then **Event**, then **Input from Simulink**. This appears as in Fig. 14.33.

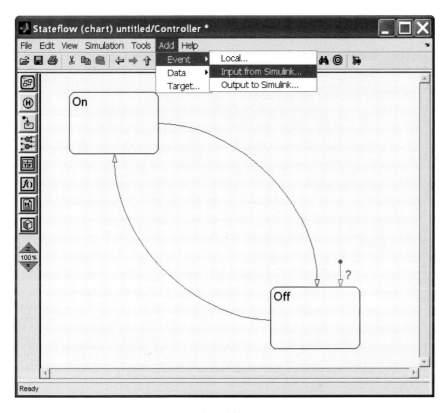

Fig. 14.33

The default trigger event is **Rising**. We wish to change this to **Either**, so that any manual change to the switch setting will toggle the active state. This is accomplished from the **Trigger** pull-down menu. Refer to Figs 14.34 and 14.35.

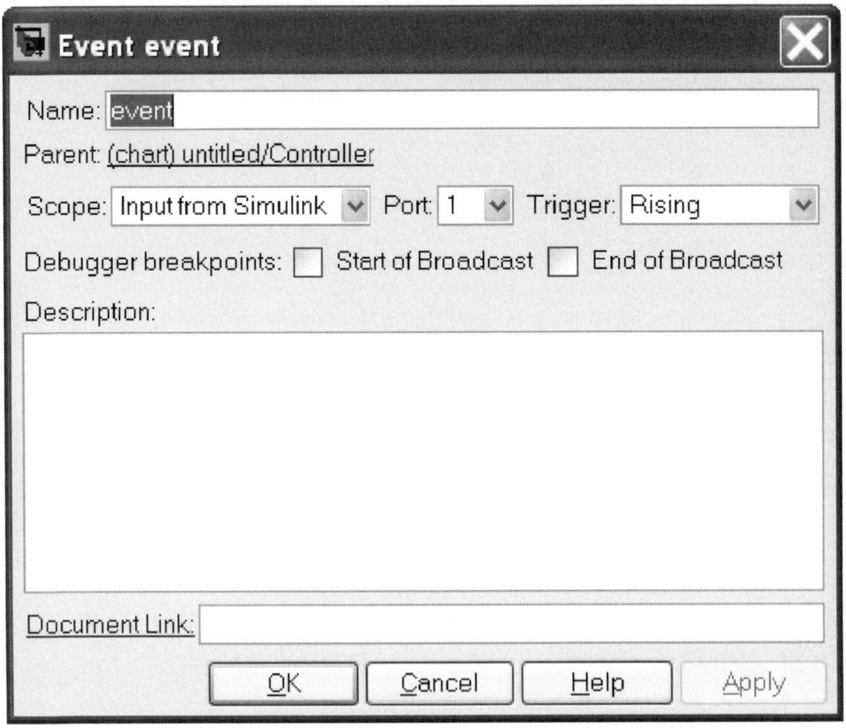

Fig. 14.34

Next select **Either** as the **Trigger** type.

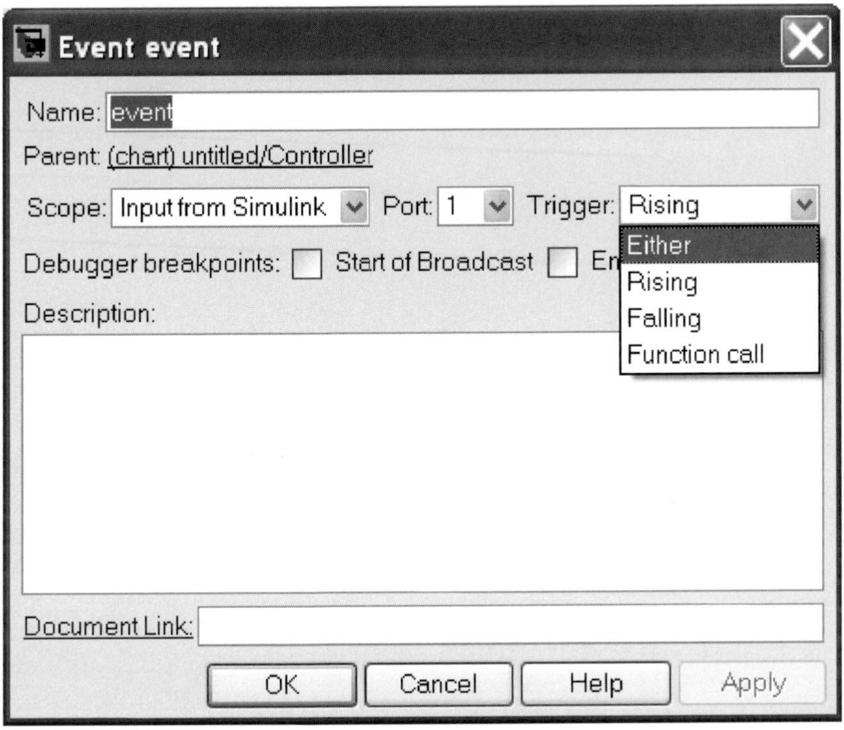

Fig. 14.35

Select **Apply** and **OK** to complete this action (see Fig. 14.36).

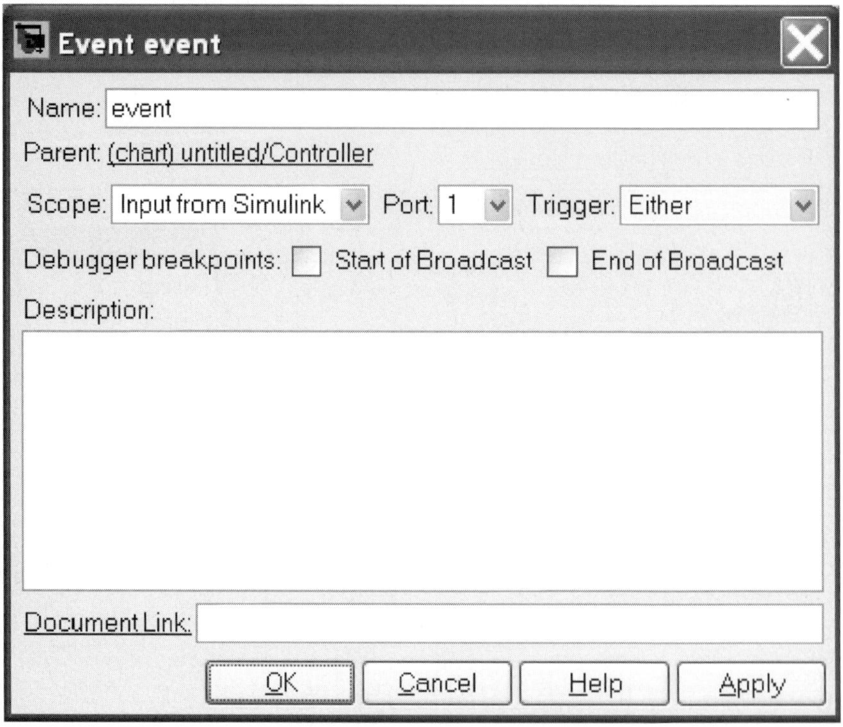

Fig. 14.36

The **Controller** block will now contain a **Trigger** input as shown in the upper center of Fig. 14.37.

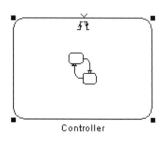

Controller

Fig. 14.37

We will use a constant value of **1** into the **Trigger** to set the active state to **On** and a value of **−1** to set the active state to **Off** within the **Controller**. Therefore, we next need to change **Constant1** from its default value of **1** to a value of **−1**. This is accomplished by double-clicking on the **Constant1** block or by selecting this block and then using the right mouse button to bring up the menu and choose **Constant Parameters...** from the resulting list. Finally, change the **Constant value** from **1** to **−1** and select **OK** (see Fig. 14.38).

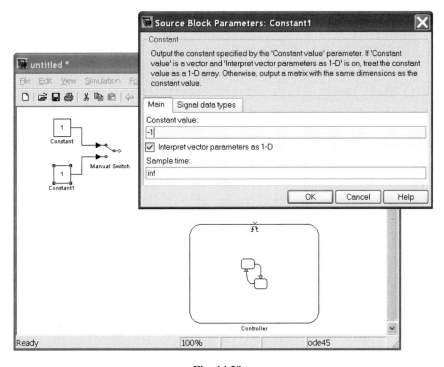

Fig. 14.38

We now need to connect the **Manual Switch**'s output to the **Trigger** on the **Controller** block. This operation is accomplished as we have done previously with Simulink only diagrams. Select the **Manual Switch**; then while still holding down the left mouse button, drag the signal path to the **Trigger** input. This operation appears as in Fig. 14.39.

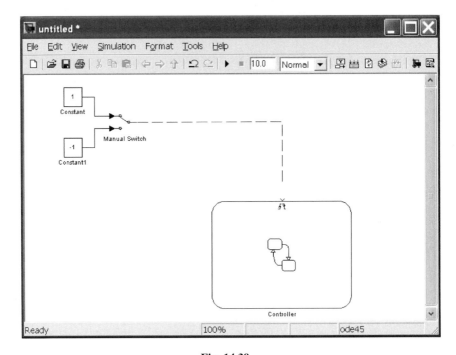

Fig. 14.39

When this operation is completed, the dashed line will turn into a solid line and will end in an arrowhead. The trigger signal will now drive the states within the **Controller Chart**.

As we have nearly completed this diagram, we should save it within our **work** folder. Either click on the **disk** icon, or go to the **File**, **Save As** option from the pull-down menu (see Fig. 14.40).

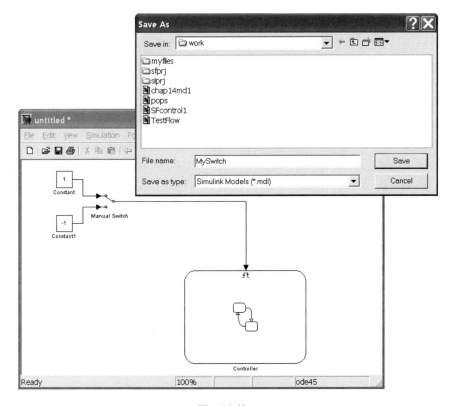

Fig. 14.40

Note that the simulation is still set to the default finish time of 10.0 seconds, and the default continuous time **ode45** integrator. Both of these simulation parameters will need to be modified.

First let us change the simulation end time to infinity (**inf**). The quickest way to do this is to type **inf** in the end time window in the upper toolbar as in Fig. 14.41.

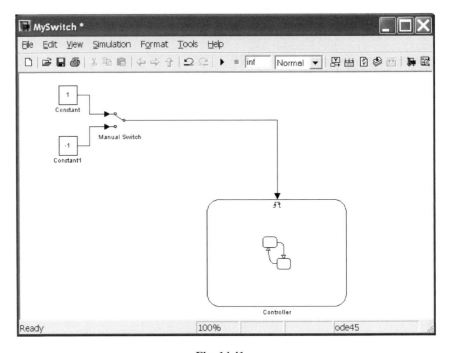

Fig. 14.41

Now we need to select the variable-step discrete-time integration algorithm. The type of integration algorithm to be used is selected from the **Simulation, Configuration Parameters...** window. The shortcut to this window is **Ctrl+E**. The following two figures show the process of generating this window and then selecting the **discrete (no continuous states)** integration algorithm. Note that the simulation end time can also be modified in the **Configuration Parameters** window (see Fig. 14.42).

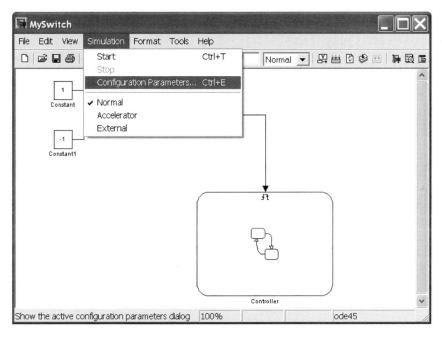

Fig. 14.42

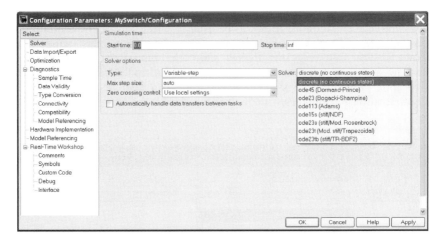

Fig. 14.43

To confirm that the correct integration algorithm is now being used, look at the lower right-hand corner of the **MySwitch** window (Fig. 14.43). It now shows that the **VariableStepDiscrete** integration algorithm is being used (Fig. 14.44).

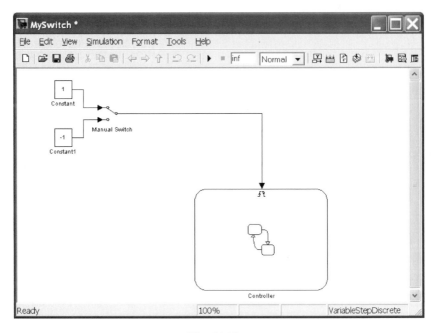

Fig. 14.44

Because the previous operations are seen as modifications to the Simulink/
Stateflow model, you need to save the model to disk again. This can be accomplished either by selecting the **disk** symbol, by using **Ctrl-S**, or by using **File**,
Save. You can then check to see that the Stateflow diagram is receiving the
Trigger signal properly by double-clicking on the **Controller** block. The Stateflow diagram then appears as in Fig. 14.45. Note that all of the transition paths are
blue, denoting proper connection paths.

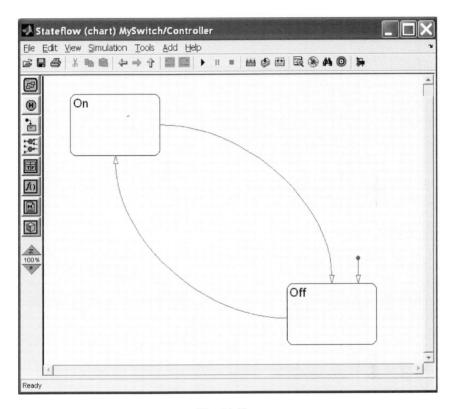

Fig. 14.45

We next execute our combined Simulink/Stateflow model. This can be accomplished using the triangular **Start simulation** icon from either the Simulink or Stateflow toolbar. Once execution starts, the time will increment within the bottom right-hand corner of the Simulink diagram, as shown in Fig. 14.46.

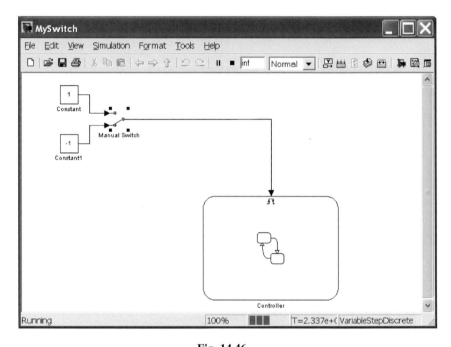

Fig. 14.46

Note that the message **Running** is also displayed in the lower left corner of the window.

Now we will examine the behavior of the Stateflow model while it is executing. Double-click on the **Manual Switch** icon, as was done in Fig. 14.46. This will change the value sent to the **Controller** via the **Trigger** from a value of **1** to a value of **−1**. This will set the active state to **Off** and thus will change this state's black outline into a bold blue outline. **Off** will also change from black to blue (see Fig. 14.47).

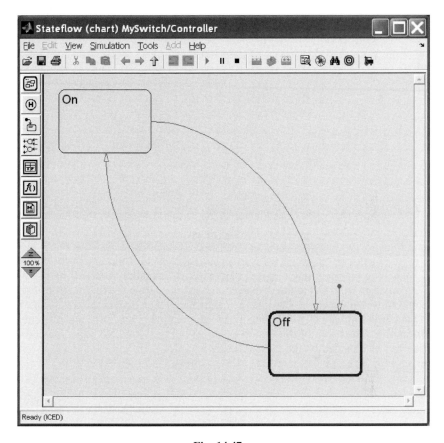

Fig. 14.47

Again, double-click on the **Manual Switch** in the Simulink diagram. You will see that the switch changes back to input the **Constant** value of **1** into the **Trigger** contained within the **Controller** Stateflow chart. This is shown in the Simulink figure (Fig. 14.48).

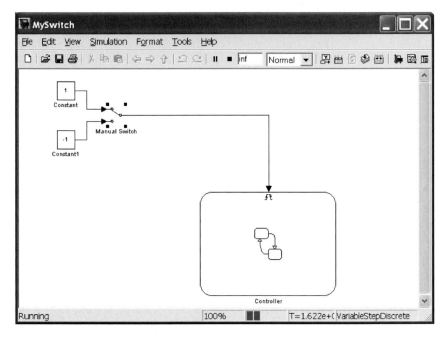

Fig. 14.48

Watching the Stateflow diagram **MySwitch/Controller**, you will briefly see the **transition** from **Off** to **On** displayed in bold blue as the active state transitions from **Off** to **On**. Once the **On** state is reached, it is then activated. This is shown in Stateflow software by displaying the block's outline in bold blue and the letters **On** in blue. The result of this action is shown in Fig. 14.49. Note that the colors used can be changed by the user as desired.

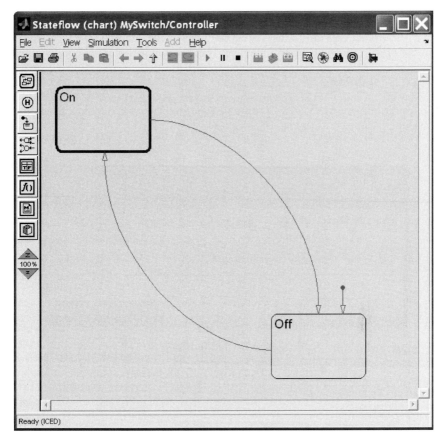

Fig. 14.49

This concludes our modeling of a switch within Stateflow software.

14.4 Using a Stateflow® Truth Table

The other element in the Stateflow library is the **Truth Table**. We will next construct a model using this element. If the Stateflow library is not open, type

>> **stateflow**

from within the MATLAB **Command Window**. Then the Stateflow block library **sflib** will appear as in Fig. 14.50.

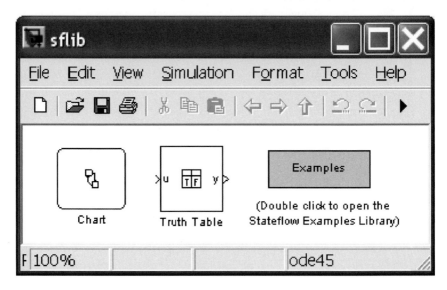

Fig. 14.50

The **Truth Table** generator is provided in the center of this window. To start, create a new Simulink model and drag a copy of the Stateflow **Truth Table** block into this model's window. This can be done from **sflib** by going to **File**, **New**, **Model** as shown in Fig. 14.51.

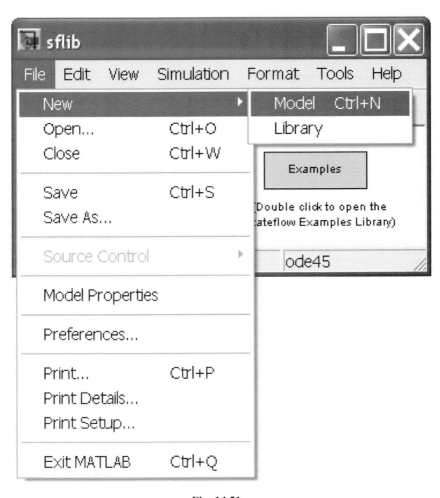

Fig. 14.51

An untitled Simulink model window appears as in Fig. 14.52.

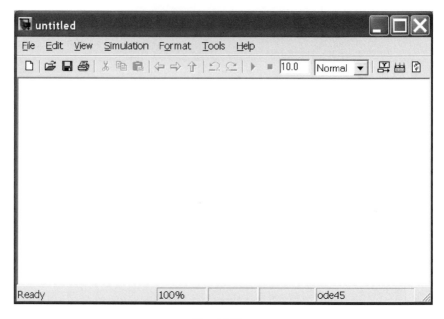

Fig. 14.52

Next drag a copy of the Stateflow block **Truth Table** into the **untitled** model window as in Fig. 14.53. Note the asterisk * after **untitled** in the **Title bar**, denoting an unsaved Simulink model.

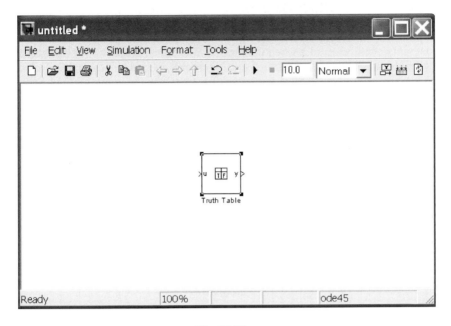

Fig. 14.53

Next open the **Simulink Library Browser** from the MATLAB **Command Library**. This can either be accomplished by typing **simulink** in the MATLAB **Command Window** or by double-clicking on the Simulink icon in the upper toolbar.

We will now construct a model similar to the one in the previous section. An input value (**u**) of **1** will be used to set the **Truth Table** to a value of **T** (true). An input value of **−1** will be used to set the **Truth Table** to a value of **F** (false). The output (**y**) will be set to **1** when the input is **1**, and the output (**y**) will be set to **0** when the input is **−1**. We will use two **Constant** blocks and a **Manual Switch** to create the two possible inputs values as was done in the previous section. These details are omitted here. The resulting Simulink model appears as in Fig. 14.54.

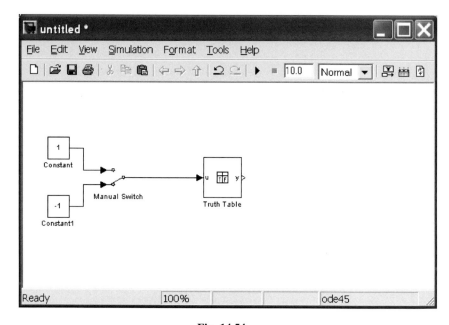

Fig. 14.54

Save this model under the name **MyTruth**. This can be accomplished by clicking on the **Save** (disk) icon or by using the **File, Save As...** pull-down menu (see Fig. 14.55).

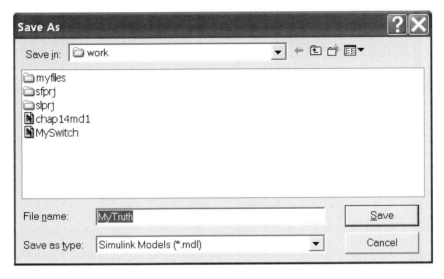

Fig. 14.55

Next open the **Configuration Parameters** window using either the **Simulation** pull-down menu or the **Ctrl-E** shortcut. Set the simulation **Stop time** to **inf**, and select the **discrete (no continuous states)** equation **Solver**. When these modifications are complete, **Apply** the changes and approve them using **OK**. The **Configuration Parameters** will now appear as in Fig. 14.56.

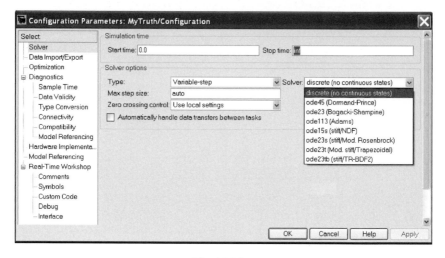

Fig. 14.56

A sink is required to display the output from the **Truth Table**. Select a **Display** block from the Simulink **Sources** library and drag a copy onto the right-hand side of the Simulink model **MyTruth**. This model will now appear as in Fig. 14.57.

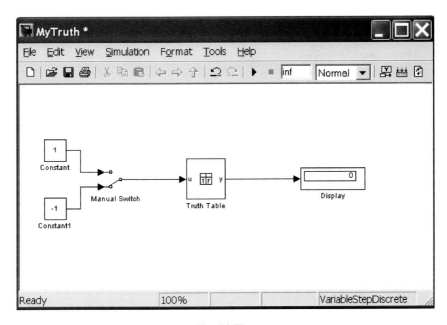

Fig. 14.57

The **Truth Table** needs to be completed next. Opening the **Truth Table** provides the user with the example shown in Fig. 14.58.

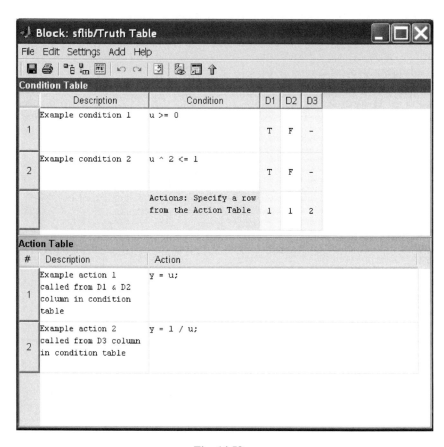

Fig. 14.58

To provide the equivalent **Truth Table** to the model from the previous section, we will modify it as follows. We only have one **Condition u ≥ 0**. When **u ≥ 0**, **Decision 1 (D1)** is set to true **(T)**. When **u ≥ 0** is false, **Decision 2 (D2)** is set to false **(F)**. As **Condition** number **2** is not needed, simply click on the **2** box and hit the **Delete** key to remove it. Next, because **Decision 3** is not needed, simply click on the **D3** box and hit the **Delete** key to remove it.

We will make one of two decisions depending on the truth of the **Condition**. If the **Condition** is true **(T)**, then **D1** will specify **Action #1** from the **Action Table**. In this case, **y = u**, and the output **(y)** will be set to **1**, denoting that the switch is on. If the **Condition** is false **(F)**, then **D2** will specify **Action #2** from the **Action Table**. In this case, **y = 0**, and the output will be set to **0**, denoting that the switch is off. This is shown in Fig. 14.59.

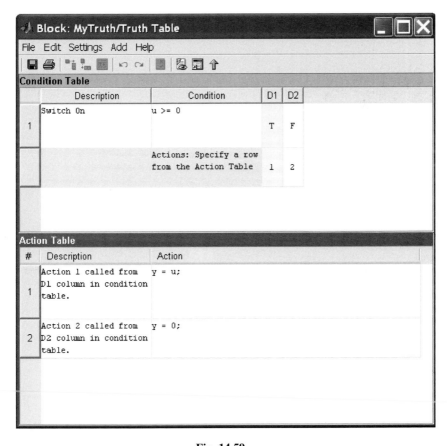

Fig. 14.59

Note that if additional conditions are needed, the **Append Row** icon ⁺ᵢ can be clicked, or the **Edit**, **Append Row** pull-down menu can be used. Similarly, if additional decisions are needed, the **Append Column** icon ⁺ can be clicked, or the **Edit**, **Append Column** pull-down menu can be used.

Our combined Simulink/Stateflow model using a **Truth Table** is shown in Fig. 14.60. This model is executed using the **Start Simulation** icon; the **Simulation**, **Start** pull-down menu; or the **Ctrl+T** shortcut. The **Display** shows a **0**, because the **Manual Switch** is set to the off position, which is equivalent to selecting as an input a constant value of −**1**.

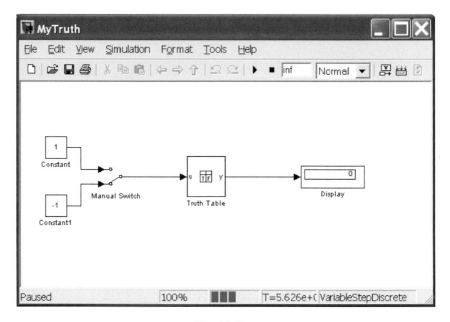

Fig. 14.60

Now double-click on the **Manual Switch**. This will change the position of the switch, so that the input value is changed to a constant value of **1**. This will set the switch to the on position, denoted by an output value of **1**. This value is shown in **Display** in Fig. 14.61.

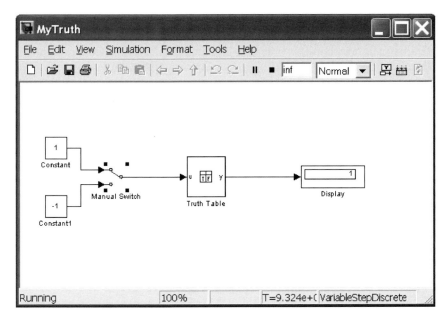

Fig. 14.61

This concludes this example of Stateflow **Truth Tables**.

14.5 Conclusion

This chapter concludes our discussion of Stateflow software and our tutorials on MATLAB, Simulink, and Stateflow products. You are now ready to further explore the capabilities of these powerful tools on your own!

Practice Exercises

14.1   The following exercises are provided to give the reader experience in building and executing a combined Simulink/Stateflow model. You will use the Stateflow **Chart** block to include a Stateflow diagram within a Simulink model. The model must be constructed to accomplish the equivalent to the following MATLAB statement:

if cond <= 36
    output = cond^2;
elseif cond > 36 & cond <= 216
    output = cond^3;
else
    output = cond;
end

Use Fig. 14.62 to assist you in constructing your Simulink diagram, which will include a Stateflow **Chart** block.

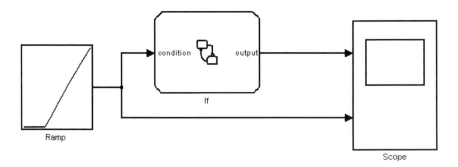

Fig. 14.62

The Stateflow block **Chart** will be constructed to accomplish the equivalent to the previously provided MATLAB statement as in Fig. 14.63.

This model is constructed using the elements on the left-hand side of the Stateflow window. After constructing this model, simulate it using a **Ramp** as a **Source** to verify that the **Chart** is functioning as in the MATLAB statement.

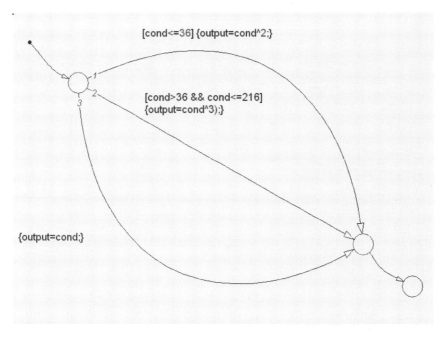

Fig. 14.63

Notes

Appendix A
History of MATLAB® and The MathWorks, Inc.

The history of MATLAB® starts with the development of two libraries of FORTRAN mathematical subroutines in the mid-1970s under a grant from the National Science Foundation. These two libraries were called LINPACK and EISPACK. LINPACK was a software library for performing numerical linear algebra. LINPACK made use of the BLAS (Basic Linear Algebra Subprograms) libraries for performing basic vector and matrix operations. EISPACK was a software library for solving eigenvalue problems and conducting related analyses. LINPACK was written by Jack Dongarra, Jim Bunch, Cleve Moler, and Pete Stewart. EISPACK was developed by several authors located primarily at Argonne National Laboratory, a group that included Dongarra and Moler. The two libraries have been superseded by LAPACK, whose routines run more efficiently on present-day computer architectures.

Short for "MATrix LABoratory," MATLAB was developed in the late 1970s by Cleve Moler. The software has been identified by the name MATLAB since March 1979. Moler was then chairman of the computer science department at the University of New Mexico. Moler taught mathematics and computer science for almost 20 years at the University of New Mexico. He also worked at the University of Michigan and spent sabbatical leaves at Stanford University. The software was a personal project to provide his students with access to LINPACK and EISPACK in his linear algebra courses without having to program in FORTRAN and to take advantage of recently developed interactive computer system capabilities.

MATLAB soon spread to other universities and found a strong audience within the applied mathematics community. Moler would provide copies for use at other universities and institutions, either after giving a talk at that location or upon request. All he would request was $75 to reimburse his cost for the two nine-track tapes and the shipping costs.

Others saw the commercial potential of MATLAB and offered upgraded versions of the product. Two of these products, which were available before The MathWorks was founded, were Matrix-X and Ctrl-C. Matrix-X initially was available only for mainframe computers such as IBM and DEC systems. It added graphics and control system design capabilities to MATLAB's analysis capabilities. It was made available on UNIX workstations, such as those by Apollo, in the mid-1980s. Ctrl-C was a similar, competitive program based on MATLAB available on DEC computer systems such as the VAX-780 series of

425

computers. Jack Little was involved in the development of the Ctrl-C product. The final competitor to MATLAB at the time, Easy-5, was not based on MATLAB but on dynamics software developed by Boeing.

In early 1980, when Moler visited Stanford University, Little was first exposed to MATLAB. Little, an engineer, recognized the potential application of MATLAB to engineering applications on the relatively recently developed personal computers (PCs). In 1983, Little, Moler, and Steve Bangert worked to develop a second-generation, professional version of MATLAB with graphics capabilities. Little was also involved in the development of the Signal Processing Toolbox. The MathWorks, Inc., was founded in 1984 to market and continue development of MATLAB. The headquarters is now located in Natick, Massachusetts.

Later, Matrix-X was rewritten in C, and its syntax was changed from that used in MATLAB. Until then, MATLAB commands and programs worked with few exceptions in Matrix-X and Ctrl-C. Little worked with Moler and Bangert to rewrite MATLAB into C in the mid-1980s. These rewritten libraries were then known as the Control System Toolbox. Little's specialty was as a control design engineer. This was the first MATLAB toolbox offered by The MathWorks.

There are now two basic versions of MATLAB: the professional version and the student edition. For a while the student edition was distributed by Prentice-Hall; now both are again distributed by The MathWorks, Inc.

Later in the 1980s, graphical modeling environments were developed for these mathematical software tools. Grumman Aircraft developed Protoblock, a graphical system for the nonlinear modeling and simulation of dynamic systems in MATLAB. This system was available as a third-party product for MATLAB until Northrop acquired Grumman and the product line was dropped. Another third-party graphical interface was briefly available for MATLAB on IBM PCs in the late 1980s. Integrated Systems, Inc., which marketed and supported Matrix-X, developed the System Build graphical system for its product. That system was mostly offered on UNIX workstations but was ported to other computer systems. Boeing developed a graphical modeling system for Easy-5. ACSL, a nonlinear simulation based on the older IBM CSMP simulation environment, was offered with the EASE graphical system.

Simulink® is the graphical system for the nonlinear modeling and simulation of dynamic systems within the MATLAB environment. Originally, it was called SIMULAB, but the name was already copyrighted. Simulink reached the market in 1990. The principal authors were Joseph Hicklin, who wrote the user interface, and Andrew Grace, who wrote the numerical routines. Simulink is a graphical, mouse-driven program that allows systems to be modeled by drawing a block diagram on the screen. It can handle linear, nonlinear, continuous-time, discrete-time, multivariable, and multirate systems. It was designed to take full advantage of windowing technology, including pull-down menus and mouse interactions.

Object or state modeling environments started to be seen in the mid-1990s. Object-time was an early such environment. Stateflow® software was first a separate environment from Simulink. Stateflow and Simulink models could not be used together in the first release of Stateflow software. The second version of Stateflow software integrated these products.

The MathWorks, Inc., is still run by Little, who is the company's president and CEO. Moler is the company's chairman and chief scientist. The MathWorks is a privately held company. The over 1,300 employees of The MathWorks refer to themselves as MathWorkers. The company develops and markets an extensive family of add-on products to meet the specific needs of the scientific, engineering, and financial communities. Over one million people and 3,500 universities use MATLAB and its related products in over 100 countries. The MathWorks has been profitable every year since its inception.

The list of MATLAB, Simulink, and Stateflow releases follows:

MATLAB—The University of New Mexico
MATLAB 1.3
MATLAB 2.0
MATLAB 2.4
MATLAB 3.0
MATLAB 3.5
MATLAB 4.0
MATLAB 4.1
MATLAB 4.2
MATLAB 5.0 (R8)
MATLAB 5.1 (R9)
MATLAB 5.2 (R10)
MATLAB 5.3 (R11)
MATLAB 5.3.1 (R11.1)
MATLAB 6.0 (R12)
MATLAB 6.1 (R12.1)
MATLAB 6.5 (R13)
MATLAB 6.5.1 (R13SP1)
MATLAB 6.5.2 (R13SP2)
MATLAB 7.0 (R14)
MATLAB 7.0.1 (R14SP1)
MATLAB 7.0.4 (R14SP2)
MATLAB 7.1 (R14SP3)
MATLAB 7.2 (R2006a)
MATLAB 7.3 (R2006b)

SIMULINK—Development of the product
Simulink 1.1
Simulink 1.2
Simulink 1.3
Simulink 2.0 (R8)
Simulink 2.1 (R9)
Simulink 2.2 (R10)
Simulink 3.0 (R11)
Simulink 3.0.1 (R11.1)
Simulink 4.0 (R12)
Simulink 4.1 (R12.1)
Simulink 5.0 (R13)

Simulink 5.1 (R13SP1)
Simulink 5.2 (R13SP2)
Simulink 6.0 (R14)
Simulink 6.1 (R14SP1)
Simulink 6.2 (R14SP2)
Simulink 6.3 (R14SP3)
Simulink 6.4 (R2006a)
Simulink 6.5 (R2006b)

Stateflow 1.0 (R9)
Stateflow 1.0.6 (R10)
Stateflow 2.0 (R11)
Stateflow 2.0.1 (R11.1)
Stateflow 3.0.2 (R11.1+)
Stateflow 4.0 (R12)
Stateflow 4.1 (R12.1)
Stateflow 4.1 (R13)
Stateflow 5.1 (R13+)
Stateflow 5.1.1 (R13SP1)
Stateflow 5.1.2 (R13SP2)
Stateflow 6.0 (R14)
Stateflow 6.1 (R14SP1)
Stateflow 6.2 (R14SP2)
Stateflow 6.3 (R14SP3)
Stateflow 6.4 (R2006a)
Stateflow 6.5 (R2006b)

Appendix B
Tuning MATLAB®, Simulink®, and Stateflow® Solvers

B.1 Improving Simulation Performance and Accuracy

Simulation performance and accuracy within MATLAB®, Simulink®, and Stateflow® software can be affected by many things, including the model design and choice of configuration parameters. The MATLAB, Simulink, and Stateflow solvers handle most model simulations accurately and efficiently using their default parameter values. However, some models yield better results if you adjust the solver's parameters. Also, providing information about your model's behavior to the solver can improve your simulation results.

B.1.1 Speeding Up Simulations

Slow simulation speed in MATLAB, Simulink, or Stateflow software can have many causes. A few of these follow. First, your model includes a **MATLAB Fcn** block. When a model includes a **MATLAB Fcn** block, the MATLAB interpreter is called at each time step, drastically slowing down the simulation. To improve the speed of your simulation, use a built-in **MATLAB Fcn** block or **Math Function** block whenever possible. Your simulations can also be slowed if your model includes an M-file **S-Function**. M-file **S-Functions** also cause the MATLAB interpreter to be called at each time step. Consider either converting the S-Function to a subsystem or to a C-MEX file **S-Function**. Another thing that can slow your model is if it includes a **Memory** block. Using a **Memory** block causes the variable-order solvers (**ode15s** and **ode113**) to be reset back to order 1 at each time step.

Do not set the maximum step size to be too small. If you changed the maximum step size, try running the simulation again with the default value (**auto**). Do not ask for too much accuracy. The default relative tolerance (0.1% accuracy) is usually sufficient. For models with states that go to zero, if the absolute tolerance parameter is too small, the simulation can take too many steps around the near-zero state values. See the discussion of error under **Maximum order** in the **Help** window. If the time scale is too long, reduce the time interval.

Problems can arise if the system is stiff but you are using a nonstiff solver. Try using **ode15s**. If the model uses sample times that are not multiples of each other, the solver is forced to take small enough steps to ensure sample time hits for all sample times. Your model can also be slowed if it contains an algebraic loop. The solutions to algebraic loops are iteratively computed at every time step. Therefore, they severely degrade performance. For more information, look up **Algebraic Loops** under **Help**. Simulink models can be slowed down if a **Random Number** block is fed into an **Integrator** block. For continuous systems, use the **Band-Limited White Noise** block in the **Sources** library

B.1.2 Improving Simulation Accuracy

To check your simulation's accuracy, run the simulation over a reasonable time span. Then either reduce the relative tolerance to 1e-4 (the default is 1e-3), or reduce the absolute tolerance, and then run your simulation again. Next compare the results of both simulation runs. If the results are not significantly different, you can feel confident that the solution has converged. If the simulation misses significant behavior at its start, reduce the initial step size to ensure that the simulation does not step over the missed significant behavior.

If the simulation results become unstable over time, your system might simply be unstable! If you are using **ode15s**, you might need to restrict the maximum order to two (the maximum order for which the solver is A-stable) or try using the **ode23s** solver.

Multiple approaches can be tried if the simulation results do not appear to be accurate. For a model that has states whose values approach zero, if the absolute tolerance parameter is too large, the simulation can take too few steps around areas of near-zero state values. Reduce the parameter value or adjust it for individual states using the **Integrator**'s dialog box. If reducing the absolute tolerances does not sufficiently improve the accuracy, reduce the size of the relative tolerance parameter to reduce the acceptable error and to force smaller step sizes and more steps.

One of the most commonly used simulation commands is the **sim** function. Note that it accepts linear plant models only. If your plant is a nonlinear Simulink model, you could control it as demonstrated in **Nonlinear Plants** (see **Help**). The full syntax of the command that runs the simulation is

[t,x,y] = sim(model, timespan, options, ut);

Only the model parameter is required in the previous command. Parameters not supplied within the command are taken from the **Configuration Parameters** dialog box settings.

For detailed syntax on the **sim** command, see the **Help** documentation for the **sim** command. The options parameter is a structure that supplies additional configuration parameters, including the solver name and error tolerances. You can define the parameters in the options structure using the **simset** command.

B.1.3 Simulation Commands

The following are the available simulation commands: **add_exec_event_ listener**, **model**, **sim**, **simplot**, **simset**, **simget**, and **slbuild**.

To follow are some of the important factors in efficiently executing simulations.

B.1.3.1 Absolute error tolerance

AbsTol   default - positive scalar {1e-6}

This scalar applies to all elements of the state vector. **AbsTol** applies only to the variable-step solvers.

B.1.3.2 Relative error tolerance

RelTol   default - positive scalar {1e-3}

This property applies to all elements of the state vector. The estimated error in each integration step satisfies

e(i) <=max(RelTol*abs(x(i)),AbsTol(i))

This property applies only to the variable-step solvers and defaults to 1e-3, which corresponds to accuracy within 0.1%.

B.1.3.3 Tracing facilities.
This property enables simulation tracing facilities. You specify one or more as a comma-separated list. The **minstep** trace flag specifies that the simulation stops when the solution changes so abruptly that the variable-step solvers cannot take a step and satisfy the error tolerances. By default, Simulink issues a warning message and continues the simulation. The **siminfo** trace flag provides a short summary of the simulation parameters in effect at the start of simulation. The **compile** trace flag displays the compilation phases of a block diagram model.

B.1.3.4 ZeroCross.
This command enables (default) or disables the location of the zero crossings. This property applies only to the variable-step solvers. If set to off, variable-step solvers do not detect zero crossings for blocks having intrinsic zero-crossing detection. Then the solvers adjust their step sizes only to satisfy the error tolerance.

B.1.3.5 Debug.
The default for **debug** is **off**. Setting this to **on** starts the simulation in debug mode (see **Starting the Debugger** in the on-line Simulink **Help** documentation for more information). The value of this option can be a cell array of commands to be sent to the debugger after it starts, e.g.,

```
opts = simset('debug', ...
         {'strace 4', ...
         'diary solvertrace.txt', ...
         'cont', ...
         'diary off', ...
         'cont'})

sim('vdp',[], opts);
```

B.2   Selecting Solvers

When running the simulation, Simulink solves the dynamic system using one of several solvers. You can specify several solver options using the **Solver Options** panel in the **Options** dialog box. The type of solver can be variable step or fixed step. Variable-step solvers keep the error within specified tolerances by adjusting the step size the solver uses. Fixed-step solvers use a constant step size. When your model's states are likely to vary rapidly, a variable-step solver is often faster.

B.2.1   Variable-Step Solvers

When you select **Variable-step** as the solver type, you can choose any of the solvers and integration techniques listed in Table B.1.

See the Simulink **Help** documentation for information on these solvers.

When you select **Variable-step** as the solver **Type**, you can also set several other parameters that affect the step size of the simulation. These include first the **Maximum step size**, the largest step size Simulink can use during a simulation. The **Minimum step size** is the smallest step size Simulink can use during a simulation. The **Initial step size** is the step size Simulink uses to begin the simulation. The **Relative tolerance** is the largest allowable relative error at any step in the simulation. The **Absolute tolerance** is the largest allowable absolute error at any step in the simulation. The **Zero crossing control** needs to be set to **on** for the solver to compute exactly where the signal crosses the x axis. This is useful when using functions that are nonsmooth and when the output depends on when a signal crosses the x axis. An example of this would be the use of absolute values.

By default, Simulink automatically chooses values for these options. To choose your own values, enter them into the appropriate fields. For more information on these options and the circumstances in which to use them, see the Simulink **Help** documentation.

B.2.2   Fixed-Step Discrete Solvers

When the **Type** control of the **Solver** configuration pane is set to fixed step, the configuration pane's solver control allows you to choose one of the set of

Table B.1   Variable-step solvers

Solver	Integration technique
Discrete	No continuous states
ode45	Dormand-Prince
ode23	Bogacki-Shampine
ode113	Adams
ode15s	Stiff/NDF
ode23s	Stiff/mod. Rosenbrock
ode23t	Mod. Stiff/trapezoidal
ode23tb	Stiff/TR-BDF2

fixed-step solvers that Simulink provides. The set of fixed-step solvers comprises two types of solvers: discrete and continuous.

B.2.2.1 Choosing a fixed-step discrete solver.

The fixed-step discrete solver computes the time of the next time step by adding a fixed step size to the time of the current time. The accuracy and length of time of the resulting simulation depends on the size of the steps taken by the simulation. The smaller the step size is, the more accurate the results are, but the longer the simulation takes. You can allow Simulink to choose the size of the step size (the default), or you can choose the step size yourself. If you allow Simulink to choose the step size, Simulink sets the step size to the fundamental sample time of the model if the model has discrete states or to the result of dividing the difference between the simulation's start and stop time by 50 if the model has no discrete states. This choice assures that the simulation will hit every simulation time required to update the model's discrete states at the model's specified sample times

B.2.2.2 Fixed-step discrete solver limitations.

The fixed-step discrete solver has a fundamental limitation. It cannot be used to simulate models that have continuous states. That is because the fixed-step discrete solver relies on a model's blocks to compute the values of the states that they define. Blocks that define discrete states compute the values of those states at each time step taken by the solver. Blocks that define continuous states, on the other hand, rely on the solver to compute the states. Continuous solvers perform this task. You should thus select a continuous solver if your model contains continuous states.

Note that if you attempt to use the fixed-step discrete solver to update or simulate a model that has continuous states, Simulink displays an error message. Thus, updating or simulating a model is a quick way to determine whether it has continuous states.

B.2.3 Fixed-Step Continuous Solvers

Simulink provides a set of fixed-step continuous solvers that, like the fixed-step discrete solver, compute the simulation's next time by adding a fixed-size time step to the current time. In addition, the continuous solvers employ numerical integration to compute the values of a model's continuous states at the current step from the values at the previous step and the values of the state derivatives. This allows the fixed-step continuous solvers to handle models that contain both continuous and discrete states.

Note that, in theory, a fixed-step continuous solver can handle models that contain no continuous states. However, that would impose an unnecessary computational burden on the simulation. Consequently, Simulink always uses the fixed-step discrete solver for a model that contains no states or only discrete states, even if you specify a fixed-step continuous solver for the model.

Simulink provides two distinct types of fixed-step continuous solvers: explicit and implicit solvers. Explicit solvers (see Explicit Fixed-Step Continuous Solvers under **Help**) compute the value of a state at the next time step as an explicit function of the current value of the state and the state derivative, e.g.,

$$X(n + 1) = X(n) + h * DX(n)$$

where X is the state, DX is the state derivative, and h is the step size. An implicit solver (see Implicit Fixed-Step Continuous Solvers under **Help**) computes the state at the next time step as an implicit function of the state and the state derivative at the next time step, e.g.,

$$X(n + 1) - X(n) - h * DX(n + 1) = 0$$

This type of solver requires more computation per step than an explicit solver but is also more accurate for a given step size. This solver thus can be faster than explicit fixed-step solvers for certain types of stiff systems.

B.2.3.1 Explicit fixed-step continuous solvers.
Simulink provides a set of explicit fixed-step continuous solvers. The solvers differ in the specific integration technique used to compute the model's state derivatives. Table B.2 lists the available solvers and the integration techniques they use.

The integration techniques used by the fixed-step continuous solvers trade accuracy for computational effort. Table B.2 the solvers in order of the computational complexity of the integration methods they use from the least complex (**ode1**) to the most complex (**ode5**).

As with the fixed-step discrete solver, the accuracy and length of time of a simulation driven by a fixed-step continuous solver depend on the size of the steps taken by the solver. The smaller the step size is, the more accurate the results are, but the longer the simulation takes. For any given step size, the more computationally complex the solver, the more accurate the simulation.

If you specify a fixed-step solver type for a model, Simulink sets the solver's model to **ode3**. Simulink chooses a solver capable of handling both continuous and discrete states with moderate computational effort. As with the discrete solver, Simulink by default sets the step size to the fundamental sample time of the model if the model has discrete states or to the result of dividing

Table B.2   Fixed-step solvers

Solver	Integration technique
Discrete	No continuous states
ode1	Euler's method
ode2	Heun's method
ode3	Bogacki-Shampine formula
ode4	Fourth-order Runge-Kutta (RK4) formula
ode5	Dormand-Prince formula

the difference between the simulation's start and stop time by 50 if the model has no discrete states. This assures that the solver will take a step at every simulation time required to update the model's discrete states at the model's specified sample rates. However, it does not guarantee that the default solver will accurately compute a model's continuous states or that the model cannot be simulated in less time with a less complex solver. Depending on the dynamics of your model, you may need to choose another solver and/or sample time to achieve acceptable accuracy or to shorten the simulation time.

B.2.3.2 Implicit fixed-step continuous solvers.

Simulink provides one solver in this category: **ode14x**. This solver uses a combination of Newton's method and extrapolation from the current value to compute the value of a model state at the next time step. Simulink allows you to specify the number of Newton's method iterations and the extrapolation order that the solver uses to compute the next value of a model state (see Fixed-Step Solver Options under **Help**). The more iterations and the higher the extrapolation order that you select, the greater the accuracy, but also the greater the computational burden per step size.

B.2.3.3 Choosing a fixed-step continuous solver.

Any of the fixed-step continuous solvers in Simulink can simulate a model to any desired level of accuracy, given enough time and a small enough step size. Unfortunately, in general, it is not possible, or at least not practical, to decide a priori which solver and step size combination will yield acceptable results for a model's continuous states in the shortest time. Determining the best solver for a particular model thus generally requires experimentation.

Here is the most efficient way to choose the best fixed-step solver for your model experimentally. First, use one of the variable-step solvers to simulate your model to the level of accuracy that you desire. This will give you an idea of what the simulation results should be. Next, use **ode1** to simulate your model at the default step size for your model. Compare the results of simulating your model with **ode1** with the results of simulating with the variable-step solver. If the results are the same within the specified level of accuracy, you have found the best fixed-step solver for your model, namely **ode1**. That is because **ode1** is the simplest of the Simulink fixed-step solvers and hence yields the shorted simulation time for the current step size.

If **ode1** does not give accurate results, repeat the preceding steps with the other fixed-step solvers until you find the one that gives accurate results with the least computational effort. The most efficient way to do this is to use a binary search technique. First try **ode3**. If it gives accurate results, try **ode2**. If **ode2** gives accurate results, it is the best solver for your model; otherwise, **ode3** is the best. If **ode3** does not give accurate results, try **ode5**. If **ode5** gives accurate results, try **ode4**. If **ode4** gives accurate results, select it as the solver for your model; otherwise, select **ode5**.

If **ode5** does not give accurate results, reduce the simulation step size and repeat the preceding process. Continue in this way until you find a solver that solves your model accurately with the least computational effort.

B.3   Non-Real-Time and Real-Time Simulations

The real-time program calculates the next values for the continuous states based on the derivative vector, dx/dt, for the current values of the inputs and the state vector. These derivatives are then used to calculate the next values of the states using a state-update equation. This is the state-update equation for the first-order Euler method (**ode1**):

$$x = x + (dx/dt) \, h$$

where h is the step size of the simulation, x represents the state vector, and dx/dt is the vector of derivatives. Other algorithms can make several calls to the output and derivative routines to produce more accurate estimates.

Note, however, that real-time programs use a fixed step size because it is necessary to guarantee the completion of all tasks within a given amount of time. This means that, although you should use higher-order integration methods for models with widely varying dynamics, the higher-order methods require additional computation time. In turn, the additional computation time might force you to use a larger step size, which can diminish the improvement of accuracy initially sought from the higher-order integration method.

Generally, the stiffer the equations (that is, the more dynamics in the system with widely varying time constants), the higher the order of the method that you must use. In practice, the simulation of very stiff equations is impractical for real-time purposes except at very low sample rates. You should test fixed-step-size integration in Simulink to check stability and accuracy before implementing the model for use in real-time programs.

For linear systems, it is more practical to convert the model that you are simulating to a discrete time version. For instance, use the **c2d** function in the Control System Toolbox.

Appendix C
MATLAB®, Simulink®, and Stateflow®
Quick Reference Guide

Here is a quick reference guide to frequently used MATLAB®, Simulink®, and Stateflow® commands.

Functions (Categorical List)

Basic Information
disp	Display text or array
display	Overloaded method to display text or array
isempty	Determine if input is empty matrix
isequal	Test arrays for equality
isequalwithequalnans	Test arrays for equality, treating NaNs as equal
isfloat	Determine if input is floating-point array
isinteger	Determine if input is integer array
islogical	Determine if input is logical array
isnumeric	Determine if input is numeric array
isscalar	Determine if input is scalar
issparse	Determine if input is sparse matrix
isvector	Determine if input is vector
length	Length of vector
ndims	Number of dimensions

Operators
+	Addition
+	Unary plus
−	Subtraction
−	Unary minus
*	Matrix multiplication
^	Matrix power
\	Back slash or left matrix divide
/	Slash or right matrix divide
'	Transpose
.'	Nonconjugated transpose
.*	Array multiplication (element-wise)
.^	Array power (element-wise)
.\	Left array divide (element-wise)
./	Right array divide (element-wise)

Operations and Manipulation

: (colon)	Create vectors, array subscripting, and for loop iterations
accumarray	Construct an array with accumulation
blkdiag	Block diagonal concatenation
cast	Cast variable to different data type
cat	Concatenate arrays along specified dimension
cross	Vector cross product
cumprod	Cumulative product
cumsum	Cumulative sum
diag	Diagonal matrices and diagonals of matrix
dot	Vector dot product
end	Indicate last index of array
find	Find indices of nonzero elements
fliplr	Flip matrices left-right
flipud	Flip matrices up-down
flipdim	Flip matrix along specified dimension
horzcat	Concatenate arrays horizontally
ind2sub	Multiple subscripts from linear index
ipermute	Inverse permute dimensions of multidimensional array
kron	Kronecker tensor product
max	Maximum value of array
min	Minimum value of array
permute	Rearrange dimensions of multidimensional array
prod	Product of array elements
repmat	Replicate and tile array
reshape	Reshape array
rot90	Rotate matrix 90 deg
sort	Sort array elements in ascending or descending order
sortrows	Sort rows in ascending order
sum	Sum of array elements
sqrtm	Matrix square root
sub2ind	Linear index from multiple subscripts
tril	Lower triangular part of matrix
triu	Upper triangular part of matrix
vertcat	Concatenate arrays vertically

Matrix Analysis

cond	Condition number with respect to inversion
condeig	Condition number with respect to eigenvalues
det	Determinant
norm	Matrix or vector norm
normest	Estimate matrix 2-norm
null	Null space
orth	Orthogonalization
rank	Matrix rank
rcond	Matrix reciprocal condition number estimate
rref	Reduced row echelon form
subspace	Angle between two subspaces
trace	Sum of diagonal elements

Linear Equations

\ and /	Linear equation solution
chol	Cholesky factorization
cholinc	Incomplete Cholesky factorization
cond	Condition number with respect to inversion
condest	1-norm condition number estimate
funm	Evaluate general matrix function
inv	Matrix inverse
linsolve	Solve linear systems of equations
lscov	Least-squares solution in presence of known covariance
lsqnonneg	Nonnegative least squares
lu	LU matrix factorization
luinc	Incomplete LU factorization
pinv	Moore-Penrose pseudoinverse of matrix
qr	Orthogonal-triangular decomposition
rcond	Matrix reciprocal condition number estimate

Eigenvalues and Singular Values

balance	Improve accuracy of computed eigenvalues
cdf2rdf	Convert complex diagonal form to real block diagonal form
condeig	Condition number with respect to eigenvalues
eig	Find eigenvalues and eigenvectors
eigs	Find largest eigenvalues and eigenvectors of sparse matrix
gsvd	Generalized singular-value decomposition
hess	Hessenberg form of matrix
ordeig	Eigenvalues of quasi-triangular matrices
ordqz	Reorder eigenvalues in QZ factorization
ordschur	Reorder eigenvalues in Schur factorization
poly	Polynomial with specified roots
polyeig	Polynomial eigenvalue problem
qz	QZ factorization for generalized eigenvalues
rsf2csf	Convert real Schur form to complex Schur form
schur	Schur decomposition
svd	Singular-value decomposition
svds	Singular values and vectors of sparse matrix

Matrix Logarithms and Exponentials

expm	Matrix exponential
logm	Matrix logarithm
sqrtm	Matrix square root

Factorization

balance	Diagonal scaling to improve eigenvalue accuracy
cdf2rdf	Complex diagonal form to real block diagonal form
chol	Cholesky factorization
cholinc	Incomplete Cholesky factorization
cholupdate	Rank 1 update to Cholesky factorization
lu	LU matrix factorization
luinc	Incomplete LU factorization
planerot	Givens plane rotation

qr	Orthogonal-triangular decomposition
qrdelete	Delete column or row from QR factorization
qrinsert	Insert column or row into QR factorization
qrupdate	Rank 1 update to QR factorization
qz	QZ factorization for generalized eigenvalues
rsf2csf	Real block diagonal form to complex diagonal form

Trigonometric

acos	Inverse cosine
acosd	Inverse cosine, deg
acosh	Inverse hyperbolic cosine
acot	Inverse cotangent
acotd	Inverse cotangent, deg
acoth	Inverse hyperbolic cotangent
acsc	Inverse cosecant
acscd	Inverse cosecant, deg
acsch	Inverse hyperbolic cosecant
asec	Inverse secant
asecd	Inverse secant, deg
asech	Inverse hyperbolic secant
asin	Inverse sine
asind	Inverse sine, deg
asinh	Inverse hyperbolic sine
atan	Inverse tangent
atand	Inverse tangent, deg
atanh	Inverse hyperbolic tangent
atan2	Four-quadrant inverse tangent
cos	Cosine
cosd	Cosine, deg
cosh	Hyperbolic cosine
cot	Cotangent
cotd	Cotangent, deg
coth	Hyperbolic cotangent
csc	Cosecant
cscd	Cosecant, deg
csch	Hyperbolic cosecant
hypot	Square root of sum of squares
sec	Secant
secd	Secant, deg
sech	Hyperbolic secant
sin	Sine
sind	Sine, deg
sinh	Hyperbolic sine
tan	Tangent
tand	Tangent, deg
tanh	Hyperbolic tangent

Exponential

exp	Exponential
expm1	Exponential of $x - 1$
log	Natural logarithm
log1p	Logarithm of $1 + x$

log2	Base 2 logarithm and dissect floating-point numbers into exponent and mantissa
log10	Common (base 10) logarithm
nextpow2	Next higher power of 2
pow2	Base 2 power and scale floating-point number
reallog	Natural logarithm for nonnegative real arrays
realpow	Array power for real-only output
realsqrt	Square root for nonnegative real arrays
sqrt	Square root
nthroot	Real nth root

Complex

abs	Absolute value
angle	Phase angle
complex	Construct complex data from real and imaginary parts
conj	Complex conjugate
cplxpair	Sort numbers into complex conjugate pairs
i	Imaginary unit
imag	Complex imaginary part
isreal	Determine if input is real array
j	Imaginary unit
real	Complex real part
sign	Signum
unwrap	Unwrap phase angle

Rounding and Remainder

fix	Round toward zero
floor	Round toward minus infinity
ceil	Round toward plus infinity
round	Round toward nearest integer
mod	Modulus after division
rem	Remainder after division

Discrete Math (e.g., Prime Factors)

factor	Prime factors
factorial	Factorial function
gcd	Greatest common divisor
isprime	Determine if input is prime number
lcm	Least common multiple
nchoosek	All combinations of N elements taken K at a time
perms	All possible permutations
primes	Generate list of prime numbers
rat, rats	Rational fraction approximation

Elementary Matrices and Arrays

: (colon)	Create vectors, array subscripting, and **for** loop iterations
blkdiag	Construct block diagonal matrix from input arguments
diag	Diagonal matrices and diagonals of matrix
eye	Identity matrix
freqspace	Frequency spacing for frequency response
linspace	Generate linearly spaced vectors

logspace	Generate logarithmically spaced vectors
meshgrid	Generate X and Y matrices for three-dimensional plots
ndgrid	Arrays for multidimensional functions and interpolation
ones	Create array of all ones
rand	Uniformly distributed random numbers and arrays
randn	Normally distributed random numbers and arrays
repmat	Replicate and tile array
zeros	Create array of all zeros

Specialized Matrices

compan	Companion matrix
gallery	Test matrices
hadamard	Hadamard matrix
hankel	Hankel matrix
hilb	Hilbert matrix
invhilb	Inverse of Hilbert matrix
magic	Magic square
pascal	Pascal matrix
rosser	Classic symmetric eigenvalue test problem
toeplitz	Toeplitz matrix
vander	Vandermonde matrix
wilkinson	Wilkinson's eigenvalue test matrix

Polynomials

conv	Convolution and polynomial multiplication
deconv	Deconvolution and polynomial division
poly	Polynomial with specified roots
polyder	Polynomial derivative
polyeig	Polynomial eigenvalue problem
polyfit	Polynomial curve fitting
polyint	Analytic polynomial integration
polyval	Polynomial evaluation
polyvalm	Matrix polynomial evaluation
residue	Convert between partial fraction expansion and polynomial coefficients
roots	Polynomial roots

Interpolation

dsearch	Search for nearest point
dsearchn	Multidimensional closest point search
griddata	Data gridding
griddata3	Data gridding and hypersurface fitting for three-dimensional data
griddatan	Data gridding and hypersurface fitting (dimension $> = 2$)
interp1	One-dimensional data interpolation (table look-up)
interp2	Two-dimensional data interpolation (table look-up)
interp3	Three-dimensional data interpolation (table look-up)
interpft	One-dimensional interpolation using fast Fourier transform method
interpn	Multidimensional data interpolation (table look-up)
meshgrid	Generate X and Y matrices for three-dimensional plots
mkpp	Make piecewise polynomial

ndgrid	Generate arrays for multidimensional functions and interpolation
pchip	Piecewise Cubic Hermite Interpolating Polynomial (PCHIP)
ppval	Piecewise polynomial evaluation
spline	Cubic spline data interpolation
tsearchn	Multidimensional closest simplex search
unmkpp	Piecewise polynomial details

Delaunay Triangulation and Tessellation

delaunay	Delaunay triangulation
delaunay3	Three-dimensional Delaunay tessellation
delaunayn	Multidimensional Delaunay tessellation
dsearch	Search for nearest point
dsearchn	Multidimensional closest point search
tetramesh	Tetrahedron mesh plot
trimesh	Triangular mesh plot
triplot	Two-dimensional triangular plot
trisurf	Triangular surface plot
tsearch	Search for enclosing Delaunay triangle
tsearchn	Multidimensional closest simplex search

Convex Hull

convhull	Convex hull
convhulln	Multidimensional convex hull
patch	Create patch graphics object
plot	Linear two-dimensional plot
trisurf	Triangular surface plot

Voronoi Diagrams

dsearch	Search for nearest point
patch	Create patch graphics object
plot	Linear two-dimensional plot
voronoi	Voronoi diagram
voronoin	Multidimensional Voronoi diagrams

Domain Generation

meshgrid	Generate X and Y matrices for three-dimensional plots
ndgrid	Generate arrays for multidimensional functions and interpolation

Coordinate System Conversion Cartesian

cart2sph	Transform Cartesian to spherical coordinates
cart2pol	Transform Cartesian to polar coordinates
pol2cart	Transform polar to Cartesian coordinates
sph2cart	Transform spherical to Cartesian coordinates

Ordinary Differential Equations (IVP)

ode113	Solve nonstiff differential equations, variable-order method
ode15i	Solve fully implicit differential equations, variable-order method
ode15s	Solve stiff ODEs and DAEs Index 1, variable-order method

ode23	Solve nonstiff differential equations, low-order method
ode23s	Solve stiff differential equations, low-order method
ode23t	Solve moderately stiff ODEs and DAEs Index 1, trapezoidal rule
ode23tb	Solve stiff differential equations, low-order method
ode45	Solve nonstiff differential equations, medium-order method
odextend	Extend the solution of an initial value problem
odeget	Get ODE **options** parameters
odeset	Create/alter ODE **options** structure
decic	Compute consistent initial conditions for **ode15i**
deval	Evaluate solution of differential equation problem

Delay Differential Equations

dde23	Solve delay differential equations with constant delays
ddeget	Get DDE **options** parameters
ddeset	Create/alter DDE **options** structure
deval	Evaluate solution of differential equation problem

Boundary-Value Problems

bvp4c	Solve boundary-value problems for ODEs
bvpget	Get BVP **options** parameters
bvpset	Create/alter BVP **options** structure
deval	Evaluate solution of differential equation problem

Partial Differential Equations

| pdepe | Solve initial-boundary-value problems for parabolic-elliptic PDEs |
| pdeval | Evaluates by interpolation solution computed by `pdepe` |

Optimization

fminbnd	Scalar bounded nonlinear function minimization
fminsearch	Multidimensional unconstrained nonlinear minimization, by Nelder-Mead direct search method
fzero	Scalar nonlinear zero finding
lsqnonneg	Linear least squares with nonnegativity constraints
optimset	Create or alter optimization **options** structure
optimget	Get optimization parameters from **options** structure

Numerical Integration (Quadrature)

quad	Numerically evaluate integral, adaptive Simpson quadrature (low order)
quadl	Numerically evaluate integral, adaptive Lobatto quadrature (high order)
quadv	Vectorized quadrature
dblquad	Numerically evaluate double integral
triplequad	Numerically evaluate triple integral

Specialized Math

airy	Airy functions
besselh	Bessel functions of third kind (Hankel functions)
besseli	Modified Bessel function of first kind
besselj	Bessel function of first kind

besselk	Modified Bessel function of second kind
bessely	Bessel function of second kind
beta	Beta function
betainc	Incomplete beta function
betaln	Logarithm of beta function
ellipj	Jacobi elliptic functions
ellipke	Complete elliptic integrals of first and second kind
erf	Error function
erfc	Complementary error function
erfcinv	Inverse complementary error function
erfcx	Scaled complementary error function
erfinv	Inverse error function
expint	Exponential integral
gamma	Gamma function
gammainc	Incomplete gamma function
gammaln	Logarithm of gamma function
legendre	Associated Legendre functions
psi	Psi (polygamma) function

Elementary Sparse Matrices

spdiags	Sparse matrix formed from diagonals
speye	Sparse identity matrix
sprand	Sparse uniformly distributed random matrix
sprandn	Sparse normally distributed random matrix
sprandsym	Sparse random symmetric matrix

Full to Sparse Conversion

find	Find indices of nonzero elements
full	Convert sparse matrix to full matrix
sparse	Create sparse matrix
spconvert	Import from sparse matrix external format

Working with Sparse Matrices

issparse	Determine if input is sparse matrix
nnz	Number of nonzero matrix elements
nonzeros	Nonzero matrix elements
nzmax	Amount of storage allocated for nonzero matrix elements
spalloc	Allocate space for sparse matrix
spfun	Apply function to nonzero matrix elements
spones	Replace nonzero sparse matrix elements with ones
spparms	Set parameters for sparse matrix routines
spy	Visualize sparsity pattern

Reordering Algorithms

colamd	Column approximate minimum degree permutation
colperm	Column permutation
dmperm	Dulmage-Mendelsohn permutation
randperm	Random permutation
symamd	Symmetric approximate minimum degree permutation
symrcm	Symmetric reverse Cuthill-McKee permutation

Linear Algebra

cholinc	Incomplete Cholesky factorization
condest	1-norm condition number estimate
eigs	Eigenvalues and eigenvectors of sparse matrix
luinc	Incomplete LU factorization
normest	Estimate matrix 2-norm
spaugment	Form least-squares augmented system
sprank	Structural rank
svds	Singular values and vectors of sparse matrix

Linear Equations (Iterative Methods)

bicg	BiConjugate Gradients method
bicgstab	BiConjugate Gradients Stabilized method
cgs	Conjugate Gradients Squared method
gmres	Generalized Minimum Residual method
lsqr	LSQR implementation of Conjugate Gradients on Normal Equations
minres	Minimum Residual method
pcg	Preconditioned Conjugate Gradients method
qmr	Quasi-Minimal Residual method
symmlq	Symmetric LQ method

Tree Operations

etree	Elimination tree
etreeplot	Plot elimination tree
gplot	Plot graph, as in "graph theory"
symbfact	Symbolic factorization analysis
treelayout	Lay out tree or forest
treeplot	Plot picture of tree

Math Constants

eps	Floating-point relative accuracy
i	Imaginary unit
Inf	Infinity, ∞
intmax	Largest possible value of specified integer type
intmin	Smallest possible value of specified integer type
j	Imaginary unit
NaN	Not-a-Number
pi	Ratio of a circle's circumference to its diameter, π
realmax	Largest positive floating-point number
realmin	Smallest positive floating-point number
cumprod	Cumulative product
cumsum	Cumulative sum
interp1	One-dimensional data interpolation
interp2	Two-dimensional data interpolation
prod	Product of array elements
sort	Sort array elements in ascending or descending order
sortrows	Sort rows in ascending order
sum	Sum of array elements

Correlation

corrcoef	Correlation coefficients
cov	Covariance matrix

Finite Differences and Integration

cumtrapz	Cumulative trapezoidal numerical integration
del2	Discrete Laplacian
diff	Differences and approximate derivatives
gradient	Numerical gradient
trapz	Trapezoidal numerical integration

Fourier Transforms

abs	Absolute value and complex magnitude
angle	Phase angle
cplxpair	Sort numbers into complex conjugate pairs
fft	One-dimensional discrete Fourier transform
fft2	Two-dimensional discrete Fourier transform
fftn	N-dimensional discrete Fourier Transform
fftshift	Shift DC component of discrete Fourier transform to center of spectrum
fftw	Interface to the FFTW library run-time algorithm for tuning FFTs
ifft	Inverse one-dimensional discrete Fourier transform
ifft2	Inverse two-dimensional discrete Fourier transform
ifftn	Inverse multidimensional discrete Fourier transform
ifftshift	Inverse fast Fourier transform shift
nextpow2	Next higher power of two
unwrap	Unwrap phase angle in radians

Statistics

max	Maximum elements of array
mean	Average or mean value of arrays
median	Median value of arrays
min	Minimum elements of array
mode	Most frequent value of array
std	Standard deviation
var	Variance

Time Series General Timeseries

+ − .* * ./ / .\ \	Overloaded MATLAB arithmetic operators work with time series
get (timeseries)	Query time-series property values
getdatasamplesize	Return size of data sample
getqualitydesc	Return data quality descriptions
isempty (timeseries)	Determine if timeseries object is empty
length (timeseries)	Return length of time vector
plot (timeseries)	Plot time series
set (timeseries)	Set properties of timeseries object
size (timeseries)	Size of time series
timeseries	Create timeseries object
tsdata.event	Construct time-series event object
tsprops	help tsprops provides help on time-series object properties
tstool	Start Time Series Tools GUI

Time-Series Data and Time Manipulation

addsample	Add data sample to **timeseries** object
ctranspose (timeseries)	Transpose **timeseries** object
delsample	Delete sample from **timeseries** object
detrend (timeseries)	Subtract mean or best-fit line and all NaNs from time-series data
filter (timeseries)	Shape frequency content of time-series data
getabstime (timeseries)	Extract date-string time vector into cell array
getinterpmethod	Get interpolation method for time-series object
getsampleusingtime (timeseries)	Extract specified samples into new time series
idealfilter (timeseries)	Apply ideal (noncausal) filter to time-series object
resample (timeseries)	Redefine time series based on new time vector
setabstime (timeseries)	Set times of time series as date strings
setinterpmethod	Set default interpolation method for time-series object
synchronize	Synchronize two time-series objects onto common time vector
transpose (timeseries)	Transpose time-series object
tsdateinterval	Generate uniformly spaced sequence of dates and times
vertcat (timeseries)	Vertical concatenation for time series

Time-Series Events

addevent	Add event to time series
delevent	Remove event objects from time series
gettsafteratevent	Return new time series with samples occurring at or after event
gettsafterevent	Return new time series with samples occurring after event
gettsatevent	Return new time series with samples occurring at event
gettsbeforeatevent	Return new time series with samples occurring before or at event
gettsbeforeevent	Return new time series with samples occurring before event
gettsbetweenevents	Return new time series with samples occurring between events

Time-Series Statistical

iqr (timeseries)	Interquartile range of time-series data
max (timeseries)	Maximum value of time-series data
mean (timeseries)	Mean value of time-series data
median (timeseries)	Median value of time-series data
min (timeseries)	Minimum value of time-series data
std (timeseries)	Standard deviation of time-series data
sum (timeseries)	Sum of time-series data
var (timeseries)	Variance of time-series data

Time-Series Collection General tscollection

get (tscollection)	Query time-series collection property values
isempty (tscollection)	Determine if **tscollection** is empty
length (tscollection)	Return length of time vector
plot (timeseries)	Plot time series
set (tscollection)	Set properties of **tscollection** object

size (tscollection)	Size of time-series collection
tscollection	Create time-series collection object
tstool	Open Time Series Tools GUI

Time-Series Collection Data and Time Manipulation
General tscollection

addsampletocollection	Add sample to time-series collection
addts	Add data vector or time-series object to tscollection
delsamplefromcollection	Delete sample from tscollection object
getabstime (tscollection)	Extract date-string time vector into cell array
getsampleusingtime (tscollection)	Extract specified samples into new tscollection
gettimeseriesnames	Return cell array of names of time series in tscollection object
horzcat (tscollection)	Horizontal concatenation for tscollection objects
removets	Remove time-series objects from collection
resample (tscollection)	Redefine tscollection object on new time vector
setabstime (tscollection)	Set time of time-series collection as date strings
settimeseriesnames	Change time series name
vertcat (tscollection)	Vertical concatenation for tscollection object

Programming and Data Types, Data Types Numeric

[ ]	Array constructor
arrayfun	Apply function to each element of array
cast	Cast variable to different data type
cat	Concatenate arrays along specified dimension
class	Create object or return class of object
find	Find indices and values of nonzero array elements
intmax	Largest possible value of specified integer type
intmin	Smallest possible value of specified integer type
intwarning	Control state of integer warnings
ipermute	Inverse permute dimensions of multidimensional array
isa	Determine if input is object of given class (e.g., numeric)
isequal	Test arrays for equality
isequalwithequalnans	Test arrays for equality, treating NaNs as equal
isnumeric	Determine if input is numeric array
isreal	Determine if all array elements are real numbers
isscalar	Determine if input is scalar (1-by-1)
isvector	Determine if input is vector (1-by-N or N-by-1)
permute	Rearrange dimensions of multidimensional array
realmax	Largest positive floating-point number
realmin	Smallest positive floating-point number
reshape	Reshape array
squeeze	Remove singleton dimensions from array
zeros	Create array of all zeros

Characters and Strings Description of Strings in MATLAB®

strings   MATLAB string-handling description

Creating and Manipulating Strings

blanks	Create string of space characters
char	Convert to character array (string)
cellstr	Create cell array of strings from character array
datestr	Convert date and time to string format
deblank	Strip trailing blanks from end of string
lower	Convert string to lowercase
native2unicode	Convert numeric bytes to Unicode characters
sprintf	Write formatted data to string
sscanf	Read string under format control
strcat	Concatenate strings horizontally
strjust	Justify character array
strread	Read formatted data from string
strrep	Find and replace substring
strtrim	Remove leading and trailing whitespace from string
strvcat	Concatenate strings vertically
unicode2native	Convert Unicode characters to numeric bytes
upper	Convert string to uppercase

Comparing and Searching Strings

class	Create object or return class of object
findstr	Find string within another, longer string
isa	Determine if input is object of given class (e.g., char)
iscellstr	Determine if input is cell array of strings
ischar	Determine if input is character array
isletter	Detect elements that are alphabetic letters
isscalar	Determine if input is scalar (1-by-1)
isspace	Detect elements that are ASCII white spaces
isstrprop	Determine content of each element of string
isvector	Determine if input is vector (1-by-N or N-by-1)
regexp	Match regular expression
regexpi	Match regular expression, ignoring case
regexprep	Replace string using regular expression
strcmp	Compare strings
strcmpi	Compare strings, ignoring case
strfind	Find one string within another
strmatch	Find possible matches for string
strncmp	Compare first n characters of strings
strncmpi	Compare first n characters of strings, ignoring case
strtok	Return selected parts of string

Evaluating String Expressions

eval	Execute string containing MATLAB expression
evalc	Evaluate MATLAB expression with capture
evalin	Execute MATLAB expression in specified workspace

Structures

arrayfun	Apply function to each element of array
cell2struct	Convert cell array to structure array
class	Create object or return class of object

deal	Distribute inputs to outputs
fieldnames	List field names of structure
getfield	Get field of structure array
isa	Determine if input is object of given class (e.g., struct)
isequal	Test arrays for equality
isfield	Determine if input is structure array field
isscalar	Determine if input is scalar (1-by-1)
isstruct	Determine if input is structure array
isvector	Determine if input is vector (1-by-N or N-by-1)
orderfields	Order fields of structure array
rmfield	Remove fields from structure
setfield	Set value of structure array field
struct	Create structure array
struct2cell	Convert structure to cell array
structfun	Apply function to each field of scalar structure

Cell Arrays

{ }	Construct cell array
cell	Construct cell array
cellfun	Apply function to each cell of cell array
cellstr	Create cell array of strings from character array
cell2mat	Convert cell array of matrices to single matrix
cell2struct	Convert cell array to structure array
celldisp	Display cell array contents
cellplot	Graphically display structure of cell arrays
class	Create object or return class of object
deal	Distribute inputs to outputs
isa	Determine if input is object of given class (e.g., cell)
iscell	Determine if input is cell array
iscellstr	Determine if input is cell array of strings
isequal	Test arrays for equality
isscalar	Determine if input is scalar (1-by-1)
isvector	Determine if input is vector (1-by-N or N-by-1)
mat2cell	Divide matrix into cell array of matrices
num2cell	Convert numeric array to cell array
struct2cell	Convert structure to cell array

Data Type Conversion Numeric

cast	Cast variable to different data type
double	Convert to double-precision
int8	Convert to signed 8-bit integer
int16	Convert to signed 16-bit integer
int32	Convert to signed 32-bit integer
int64	Convert to signed 64-bit integer
single	Convert to single-precision
typecast	Convert data types without changing underlying data
uint8	Convert to unsigned 8-bit integer
uint16	Convert to unsigned 16-bit integer
uint32	Convert to unsigned 32-bit integer
uint64	Convert to unsigned 64-bit integer

String to Numeric

base2dec	Convert base N number string to decimal number
bin2dec	Convert binary number string to decimal number
cast	Cast variable to different data type
hex2dec	Convert hexadecimal number string to decimal number
hex2num	Convert hexadecimal number string to double-precision number
str2double	Convert string to double-precision number
str2num	Convert string to number

Numeric to String

cast	Cast variable to different type
char	Convert to character array (string)
dec2base	Convert decimal to base N number in string
dec2bin	Convert decimal to binary number in string
dec2hex	Convert decimal to hexadecimal number in string
int2str	Convert integer to string
mat2str	Convert matrix to string
num2str	Convert number to string

Other Conversions

cell2mat	Convert cell array of matrices to single matrix
cell2struct	Convert cell array to structure array
datestr	Convert date and time to string format
func2str	Construct function name string from function handle
logical	Convert numeric values to logical
mat2cell	Divide matrix into cell array of matrices
num2cell	Convert numeric array to cell array
str2func	Construct function handle from function name string
str2mat	Form blank-padded character matrix from strings
struct2cell	Convert structure to cell array

Determine Data Type

is*	Detect state
isa	Determine if input is object of given class
iscell	Determine if input is cell array
iscellstr	Determine if input is cell array of strings
ischar	Determine if input is character array
isfield	Determine if input is character array
isfloat	Determine if input is floating-point array
isinteger	Determine if input is integer array
isjava	Determine if input is Java object
islogical	Determine if input is logical array
isnumeric	Determine if input is numeric array
isobject	Determine if input is MATLAB OOPs object
isreal	Determine if all array elements are real numbers
isstruct	Determine if input is MATLAB structure array

Arrays Array Operations

[ ]	Array constructor
,	Array row element separator
;	Array column element separator
:	Range of array elements

+	Addition or unary plus
−	Subtraction or unary minus
.*	Array multiplication
./	Array right division
.\	Array left division
.^	Array power
.'	Array (nonconjugated) transpose
arrayfun	Apply function to each element of array
end	Indicate last index of array

Basic Array Information

disp	Display text or array
display	Overloaded method to display text or array
isempty	Determine if array is empty
isequal	Test arrays for equality
isequalwithequalnans	Test arrays for equality, treating NaNs as equal
islogical	Determine if input is logical array
isnumeric	Determine if input is numeric array
isscalar	Determine if input is scalar
isvector	Determine if input is vector
length	Length of vector
ndims	Number of array dimensions
numel	Number of elements in matrix or cell array
size	Array dimensions

Array Manipulation

:	Specify range of array elements
blkdiag	Construct block diagonal matrix from input arguments
cat	Concatenate arrays along specified dimension
circshift	Shift array circularly
find	Find indices and values of nonzero elements
fliplr	Flip matrices left-right
flipud	Flip matrices up-down
flipdim	Flip array along specified dimension
horzcat	Concatenate arrays horizontally
ind2sub	Subscripts from linear index
ipermute	Inverse permute dimensions of multidimensional array
permute	Rearrange dimensions of multidimensional array
repmat	Replicate and tile array
reshape	Reshape array
rot90	Rotate matrix 90 deg
shiftdim	Shift dimensions
sort	Sort array elements in ascending or descending order
sortrows	Sort rows in ascending order
squeeze	Remove singleton dimensions
sub2ind	Single index from subscripts
vertcat	Concatenate arrays vertically

Elementary Arrays

:	Construct regularly spaced vector
blkdiag	Construct block diagonal matrix from input arguments
eye	Identity matrix
linspace	Generate linearly spaced vectors

logspace	Generate logarithmically spaced vectors
meshgrid	Generate X and Y matrices for three-dimensional plots
ndgrid	Generate arrays for multidimensional functions and interpolation
ones	Create array of all ones
rand	Uniformly distributed random numbers and arrays
randn	Normally distributed random numbers and arrays
zeros	Create array of all zeros

Operators and Operations Special Characters

:	Specify range of array elements
( )	Pass function arguments, or prioritize operations
[ ]	Construct array
{ }	Construct cell array
.	Decimal point, or structure field separator
...	Continue statement to next line
,	Array row element separator
;	Array column element separator
%	Insert comment line into code
!	Issue command to operating system
=	Assignment

Arithmetic Operations

+	Plus
−	Minus
.	Decimal point
=	Assignment
*	Matrix multiplication
/	Matrix right division
\	Matrix left division
^	Matrix power
'	Matrix transpose
.*	Array multiplication (element-wise)
./	Array right division (element-wise)
.\	Array left division (element-wise)
.^	Array power (element-wise)
.'	Array transpose

Bit-Wise Operations

bitand	Return bit-wise **and**
bitcmp	Return bit-wise complement
bitget	Get bit at specified position
bitmax	Return maximum double-precision floating-point integer
bitor	Return bit-wise **or**
bitset	Set bit at specified position
bitshift	Shift bits specified number of places
bitxor	Return bit-wise **xor**
swapbytes	Swap byte ordering

Relational Operations

| < | Less than |
| <= | Less than or equal to |

>	Greater than
>=	Greater than or equal to
==	Equal to
~=	Not equal to

Logical Operations

&&	Logical **and**
‖	Logical **or**
&	Logical **and** for arrays
|	Logical **or** for arrays
~	Logical **not**
all	Determine if all array elements are nonzero
any	Determine if any array elements are nonzero
FALSE	Return logical **0** (false)
find	Find indices and values of nonzero elements
is*	Detect state
isa	Determine if input is object of given class
iskeyword	Determine if string is MATLAB keyword
isvarname	Determine if string is valid variable name
logical	Convert numeric values to logical
TRUE	Return logical **1** (true)
xor	Logical exclusive-**or**

Set Operations

intersect	Find set intersection of two vectors
ismember	Detect members of set
setdiff	Find set difference of two vectors
issorted	Determine if set elements are in sorted order
setxor	Find set exclusive **or** of two vectors
union	Find set union of two vectors
unique	Find unique elements of vector

Date and Time Operations

addtodate	Modify date number by field
calendar	Display calendar for specified month
clock	Return current time as date vector
cputime	Return elapsed CPU time
date	Return current date string
datenum	Convert date and time to serial date number
datestr	Convert date and time to string format
datevec	Convert date and time to vector of components
eomday	Return last day of month
etime	Return time elapsed between date vectors
now	Return current date and time
tic, toc	Measure performance using stopwatch timer
weekday	Return day of week

Programming in MATLAB® M-File Functions and Scripts

( )	Pass function arguments
%	Insert comment line into code
...	Continue statement to next line
depfun	List dependencies of M-file or P-file
depdir	List dependent directories of M-file or P-file

echo	Echo M-files during execution
end	Terminate block of code
function	Declare M-file function
input	Request user input
inputname	Return variable name of function input
mfilename	Return name of currently running M-file
namelengthmax	Return maximum identifier length
nargin	Return number of function input arguments
nargout	Return number of function output arguments
nargchk	Validate number of input arguments
nargoutchk	Validate number of output arguments
pcode	Create preparsed pseudocode file (P-file)
script	Script M-file description
varargin	Accept variable number of arguments
varargout	Return variable number of arguments

Evaluation of Expressions and Functions

arrayfun	Apply function to each element of array
builtin	Execute built-in function from overloaded method
cellfun	Apply function to each cell of cell array
echo	Echo M-files during execution
eval	Execute string containing MATLAB expression
evalc	Evaluate MATLAB expression with capture
evalin	Execute MATLAB expression in specified workspace
feval	Evaluate function
iskeyword	Determine if input is MATLAB keyword
isvarname	Determine if input is valid variable name
pause	Halt execution temporarily
run	Run script that is not on current path
script	Script M-file description
structfun	Apply function to each field of scalar structure
symvar	Determine symbolic variables in expression
tic, toc	Measure performance using stopwatch timer

Timer Functions

delete	Delete timer object from memory
disp	Display information about timer object
get	Retrieve information about timer object properties
isvalid	Determine if timer object is valid
set	Display or set timer object properties
start	Start timer
startat	Start timer at specified time
stop	Stop timer
timer	Create timer object
timerfind	Return array of all visible timer objects in memory
timerfindall	Return array of all timer objects in memory
wait	Block command line until timer completes

Variables and Functions in Memory

assignin	Assign value to variable in specified workspace
datatipinfo	Produce short description of variable for debugger DataTips
genvarname	Construct valid variable name from string

global	Declare global variables
inmem	Return names of M-files, MEX-files, Java classes in memory
mislocked	Determine if M-file or MEX-file cannot be cleared from memory
mlock	Prevent clearing M-file or MEX-file from memory
munlock	Allow clearing M-file or MEX-file from memory
namelengthmax	Return maximum identifier length
pack	Consolidate workspace memory
persistent	Define persistent variable
rehash	Refresh function and file system path caches

Control Flow

break	Terminate execution of **for** or **while** loop
case	Execute block of code if condition is **true**
catch	Specify how to respond to error in **try** statement
continue	Pass control to next iteration of **for** or **while** loop
else	Conditionally execute statements
elseif	Conditionally execute statements
end	Terminate conditional block of code
error	Display error message
for	Execute block of code specified number of times
if	Conditionally execute statements
otherwise	Default part of **switch** statement
return	Return to invoking function
switch	Switch among several cases, based on expression
try	Attempt to execute block of code, and catch errors
while	Repeatedly execute statements while condition is true

Function Handles

class	Create object or return class of object
feval	Evaluate function
function_handle	Handle used in calling functions indirectly
functions	Return information about function handle
func2str	Construct function name string from function handle
isa	Determine if input is object of given class (e.g., function_handle)
isequal	Determine if function handles are equal
str2func	Construct function handle from function name string

Object-Oriented Programming MATLAB®-Classes and Objects

class	Create object or return class of object
fieldnames	List public fields belonging to object
inferiorto	Establish inferior class relationship
isa	Determine if input is object of given class
isobject	Determine if input is MATLAB OOPs object
loadobj	User-defined extension of **load** function for user objects
methods	Display information on class methods
methodsview	Display information on class methods in separate window
saveobj	User-defined extension of **save** function for user objects

subsasgn	Overloaded method for $A(I) = B$, $A\{I\} = B$, and $A.field = B$
subsindex	Overloaded method for $X(A)$
subsref	Overloaded method for $A(I)$, $A\{I\}$, and $A.field$
substruct	Create structure argument for subsasgn or subsref
superiorto	Establish superior class relationship

Java Classes and Objects

cell	Convert Java array object to cell array
class	Create object or return class of object
clear	Clear Java import list or Java class definitions
depfun	List Java classes used by M-file or P-file
exist	Determine if input is Java class
fieldnames	List public fields belonging to object
im2java	Convert image to instance of Java image object
import	Add package or class to current Java import list
inmem	Return names of M-files, MEX-files, Java classes in memory
isa	Determine if input is object of given class
isjava	Determine if input is Java object
javaaddpath	Add entries to dynamic Java class path
javaArray	Construct Java array
javachk	Generate error message based on Java feature support
javaclasspath	Set and get dynamic Java class path
javaMethod	Invoke Java method
javaObject	Construct Java object
javarmpath	Remove entries from dynamic Java class path
methods	Display information on class methods
methodsview	Display information on class methods in separate window
usejava	Determine if Java feature is supported in MATLAB
which	Display package and class name for method

Error Handling

catch	Specify how to respond to error in **try** statement
error	Display error message
ferror	Query MATLAB about errors in file input or output
intwarning	Control state of integer warnings
lasterr	Return last error message
lasterror	Last error message and related information
lastwarn	Return last warning message
rethrow	Reissue error
try	Attempt to execute block of code, and catch errors
warning	Display warning message

MEX Programming

dbmex	Enable MEX-file debugging
inmem	Return names of M-files, MEX-files, Java classes in memory
mex	Compile MEX-function from C or FORTRAN source code
mexext	Return MEX-filename extension

File I/O Filename Construction

fileparts	Return parts of filename and path
filesep	Return directory separator for platform in use
fullfile	Build full filename from parts
tempdir	Return name of system's temporary directory
tempname	Return unique string for use as temporary filename

Opening, Loading, Saving Files

importdata	Load data from various types of files
load	Load workspace variables from disk
open	Open files of various types using appropriate editor or program
save	Save workspace variables on disk
uiimport	Open Import Wizard interface to import data
winopen	Open file in appropriate application (Windows only)

Memory Mapping

disp	Display information about memory map object
get	Return memmapfile object properties
memmapfile	Construct memory map object

Low-Level File I/O

fclose	Close one or more open files
feof	Test for end-of-file
ferror	Query MATLAB about errors in file input or output
fgetl	Return next line of file as string without line terminator(s)
fgets	Return next line of file as string with line terminator(s)
fopen	Open file or obtain information about open files
fprintf	Write formatted data to file
fread	Read binary data from file
frewind	Rewind open file
fscanf	Read formatted data from file
fseek	Set file position indicator
ftell	Get file position indicator
fwrite	Write binary data to file

Text Files

csvread	Read numeric data from text file using comma delimiter
csvwrite	Write numeric data to text file using comma delimiter
dlmread	Read numeric data from text file with specified delimiter
dlmwrite	Write numeric data to text file specified delimiter
textread	Read formatted data from start of text file
textscan	Read formatted data from any point in text file

XML Documents

xmlread	Parse XML document
xmlwrite	Serialize XML Document Object Model node
xslt	Transform XML document using XSLT engine

Microsoft Excel Functions

xlsfinfo	Determine if file contains Microsoft Excel (**.xls**) spreadsheet

| xlsread | Read Microsoft Excel spreadsheet file (**.xls**) |
| xlswrite | Write Microsoft Excel spreadsheet file (**.xls**) |

Lotus123 Functions

| wk1read | Read Lotus123 WK1 spreadsheet file into matrix |
| wk1write | Write matrix to Lotus123 WK1 spreadsheet file |

Scientific Data Common Data Format (CDF)

cdfepoch	Construct cdfepoch object from date string or number
cdfinfo	Return information about CDF file
cdfread	Read CDF file
cdfwrite	Write CDF file
todatenum	Convert cdfepoch object to MATLAB datenum

Flexible Image Transport System

| fitsinfo | Return information about FITS file |
| fitsread | Read FITS file |

Hierarchical Data Format (HDF)

hdf	Interface to HDF4 files
hdfinfo	Return information about HDF4 or HDF-EOS file
hdfread	Read HDF4 file
hdftool	Start HDF4 Import Tool
hdf5	Describes HDF5 data type objects
hdf5info	Return information about HDF5 file
hdf5read	Read HDF5 file
hdf5write	Write data to file in HDF5 format

Band-Interleaved Data

| multibandread | Read band-interleaved data from file |
| multibandwrite | Write band-interleaved data to file |

Audio and Audio/Video General

audioplayer	Create audio player object
audiorecorder	Perform real-time audio capture
beep	Produce beep sound
lin2mu	Convert linear audio signal to mu-law
mmfileinfo	Information about multimedia file
mu2lin	Convert mu-law audio signal to linear
sound	Convert vector into sound
soundsc	Scale data and play as sound

SPARCstation-Specific Sound Functions

| auread | Read NeXT/SUN (**.au**) sound file |
| auwrite | Write NeXT/SUN (**.au**) sound file |

Microsoft WAVE Sound Functions

wavplay	Play sound on PC-based audio output device
wavread	Read Microsoft WAVE (**.wav**) sound file
wavrecord	Record sound using PC-based audio input device
wavwrite	Write Microsoft WAVE (**.wav**) sound file

Audio/Video Interleaved (AVI) Functions

| addframe | Add frame to AVI file |
| avifile | Create new AVI file |

aviinfo	Return information about AVI file
aviread	Read AVI file
close	Close AVI file
movie2avi	Create AVI movie from MATLAB movie

Images

exifread	Read EXIF information from JPEG and TIFF images
im2java	Convert image to instance of Java image object
imfinfo	Return information about graphics file
imread	Read image from graphics file
imwrite	Write image to graphics file

Internet Exchange URL, Zip, Tar, E-Mail

gzip	Compress files into the gzip format
gunzip	Uncompress files in the gzip format
sendmail	Send e-mail message to list of addresses
tar	Compress files into a tar-file
untar	Extract contents of a tar file
unzip	Extract contents of zip file
urlread	Read contents at URL
urlwrite	Save contents of URL to file
zip	Create compressed version of files in zip format

FTP Functions

ascii	Set FTP transfer type to ASCII
binary	Set FTP transfer type to binary
cd (ftp)	Change current directory on FTP server
close (ftp)	Close connection to FTP server
delete (ftp)	Delete file on FTP server
dir (ftp)	List contents of directory on FTP server
ftp	Connect to FTP server, creating an FTP object
mget	Download file from FTP server
mkdir (ftp)	Create new directory on FTP server
mput	Upload file or directory to FTP server
rename	Rename file on FTP server
rmdir (ftp)	Remove directory on FTP server

Graphics Basic Plots and Graphs

box	Axis box for two- and three-dimensional plots
errorbar	Plot graph with error bars
hold	Hold current graph
LineSpec	Line specification syntax
loglog	Plot using log-log scales
polar	Polar coordinate plot
plot	Plot vectors or matrices
plot3	Plot lines and points in three-dimensional space
plotyy	Plot graphs with Y tick labels on the left and right
semilogx	Semi-log scale plot
semilogy	Semi-log scale plot
subplot	Create axes in tiled positions

Plotting Tools

figurepalette	Display figure palette on figure
pan	Turn panning on or off

plotbrowser	Display plot browser on figure
plottools	Start plotting tools
propertyeditor	Display property editor on figure
zoom	Turn zooming on or off

Annotating Plots

annotation	Create annotation objects
clabel	Add contour labels to contour plot
datetick	Date formatted tick labels
gtext	Place text on two-dimensional graph using mouse
legend	Graph legend for lines and patches
texlabel	Produce the TeX format from character string
title	Titles for two- and three-dimensional plots
xlabel	X-axis labels for two- and three-dimensional plots
ylabel	Y-axis labels for two- and three-dimensional plots
zlabel	Z-axis labels for three-dimensional plots

Annotation Object Properties

arrow	Properties for annotation arrows
doublearrow	Properties for double-headed annotation arrows
ellipse	Properties for annotation ellipses
line	Properties for annotation lines
rectangle	Properties for annotation rectangles
textarrow	Properties for annotation textbox

Specialized Plotting Area, Bar, and Pie Plots

area	Area plot
bar	Vertical bar chart
barh	Horizontal bar chart
bar3	Vertical three-dimensional bar chart
bar3h	Horizontal three-dimensional bar chart
pareto	Pareto char
pie	Pie plot
pie3	Three-dimensional pie plot

Contour Plots

contour	Contour (level curves) plot
contour3	Three-dimensional contour plot
contourc	Contour computation
contourf	Filled contour plot
ezcontour	Easy-to-use contour plotter
ezcontourf	Easy-to-use filled contour plotter

Direction and Velocity Plots

comet	Comet plot
comet3	Three-dimensional comet plot
compass	Compass plot
feather	Feather plot
quiver	Quiver (or velocity) plot
quiver3	Three-dimensional quiver (or velocity) plot

Discrete Data Plots
stem	Plot discrete sequence data
stem3	Plot discrete surface data
stairs	Stairstep graph

Function Plots
ezcontour	Easy-to-use contour plotter
ezcontourf	Easy-to-use filled contour plotter
ezmesh	Easy-to-use three-dimensional mesh plotter
ezmeshc	Easy-to-use combination mesh/contour plotter
ezplot	Easy-to-use function plotter
ezplot3	Easy-to-use three-dimensional parametric curve plotter
ezpolar	Easy-to-use polar coordinate plotter
ezsurf	Easy-to-use three-dimensional colored surface plotter
ezsurfc	Easy-to-use combination surface/contour plotter
fplot	Plot a function

Histograms
hist	Plot histograms
histc	Histogram count
rose	Plot rose or angle histogram

Polygons and Surfaces
convhull	Convex hull
cylinder	Generate cylinder
delaunay	Delaunay triangulation
dsearch	Search Delaunay triangulation for nearest point
ellipsoid	Generate ellipsoid
fill	Draw filled two-dimensional polygons
fill3	Draw filled three-dimensional polygons in 3-space
inpolygon	True for points inside a polygonal region
pcolor	Pseudocolor (checkerboard) plot
polyarea	Area of polygon
ribbon	Ribbon plot
slice	Volumetric slice plot
sphere	Generate sphere
tsearch	Search for enclosing Delaunay triangle
voronoi	Voronoi diagram
waterfall	Waterfall plot

Scatter/Bubble Plots
plotmatrix	Scatter plot matrix
scatter	Scatter plot
scatter3	Three-dimensional scatter plot

Animation
frame2im	Convert movie frame to indexed image
getframe	Capture movie frame
im2frame	Convert image to movie frame
movie	Play recorded movie frames
noanimate	Change **EraseMode** of all objects to **normal**

Bit-Mapped Images

frame2im	Convert movie frame to indexed image
image	Display image object
imagesc	Scale data and display image object
imfinfo	Information about graphics file
imformats	Manage file format registry
im2frame	Convert image to movie frame
im2java	Convert image to instance of Java image object
imread	Read image from graphics file
imwrite	Write image to graphics file
ind2rgb	Convert indexed image to RGB image

Printing

frameedit	Edit print frame for Simulink and Stateflow diagram
orient	Hardcopy paper orientation
pagesetupdlg	Page setup dialog box
print	Print graph or save graph to file
printdlg	Print dialog box
printopt	Configure local printer defaults
printpreview	Preview figure to be printed
saveas	Save figure to graphic file

Handle Graphics® Finding and Identifying Graphics Objects

allchild	Find all children of specified objects
ancestor	Find ancestor of graphics object
copyobj	Make copy of graphics object and its children
delete	Delete files or graphics objects
findall	Find all graphics objects (including hidden handles)
findfigs	Display off-screen visible figure windows
findobj	Find objects with specified property values
gca	Get current Axes handle
gcbo	Return object whose callback is currently executing
gcbf	Return handle of figure containing callback object
gco	Return handle of current object
get	Get object properties
ishandle	True if value is valid object handle
set	Set object properties

Object Creation Functions

axes	Create axes object
figure	Create figure (graph) windows
hggroup	Create a group object
hgtransform	Create a group to transform
image	Create image (two-dimensional matrix)
light	Create light object (illuminates Patch and Surface)
line	Create line object (three-dimensional polylines)
patch	Create patch object (polygons)
rectangle	Create rectangle object (two-dimensional rectangle)
rootobject	List of root properties
surface	Create surface (quadrilaterals)
text	Create text object (character strings)
uicontextmenu	Create context menu (popup associated with object)

Plot Objects

areaseries	Property list
barseries	Property list
contourgroup	Property list
errorbarseries	Property list
lineseries	Property list
quivergroup	Property list
scattergroup	Property list
stairseries	Property list
stemseries	Property list
surfaceplot	Property list

Figure Windows

clc	Clear figure window
clf	Clear figure
close	Close specified window
closereq	Default close request function
drawnow	Complete any pending drawing
gcf	Get current figure handle
hgload	Load graphics object hierarchy from a FIG-file
hgsave	Save graphics object hierarchy to a FIG-file
newplot	Graphics M-file preamble for **NextPlot** property
opengl	Change automatic selection mode of OpenGL rendering
refresh	Refresh figure
saveas	Save figure or model to desired output format

Axes Operations

axis	Plot axis scaling and appearance
box	Display axes border
cla	Clear axes
gca	Get current axes handle
grid	Grid lines for two- and three-dimensional plots
ishold	Get the current hold state
makehgtform	Create a transform matrix

Operating on Object Properties

get	Get object properties
linkaxes	Synchronize limits of specified axes
linkprop	Maintain same value for corresponding properties
set	Set object properties

Three-Dimensional Visualization Surface and Mesh Plots Creating Surfaces and Meshes

hidden	Mesh hidden line removal mode
meshc	Combination mesh/contourplot
mesh	Three-dimensional mesh with reference plane
peaks	A sample function of two variables
surf	Three-dimensional shaded surface graph
surface	Create surface low-level objects
surfc	Combination surf/contourplot
surfl	Three-dimensional shaded surface with lighting
tetramesh	Tetrahedron mesh plot

trimesh	Triangular mesh plot
triplot	Two-dimensional triangular plot
trisurf	Triangular surface plot

Domain Generation

griddata	Data gridding and surface fitting
meshgrid	Generation of X and Y arrays for three-dimensional plots

Color Operations

brighten	Brighten or darken colormap
caxis	Pseudocolor axis scaling
colormapeditor	Start colormap editor
colorbar	Display color bar (color scale)
colordef	Set up color defaults
colormap	Set the color look-up table (list of colormaps)
ColorSpec	Ways to specify color
graymon	Graphics figure defaults set for gray-scale monitor
hsv2rgb	Hue-saturation-value to red-green-blue conversion
rgb2hsv	RGB to HSV conversion
rgbplot	Plot colormap
shading	Color shading mode
spinmap	Spin the colormap
surfnorm	Three-dimensional surface normals
whitebg	Change axes background color for plots

Colormaps

autumn	Shades of red and yellow colormap
bone	Gray scale with a tinge of blue colormap
contrast	Gray colormap to enhance image contrast
cool	Shades of cyan and magenta colormap
copper	Linear copper-tone colormap
flag	Alternating red, white, blue, and black colormap
gray	Linear gray-scale colormap
hot	Black-red-yellow-white colormap
hsv	Hue-saturation-value (HSV) colormap
jet	Variant of HSV
lines	Line color colormap
prism	Colormap of prism colors
spring	Shades of magenta and yellow colormap
summer	Shades of green and yellow colormap
winter	Shades of blue and green colormap

View Control Controlling the Camera Viewpoint

camdolly	Move camera position and target
camlookat	View specific objects
camorbit	Orbit about camera target
campan	Rotate camera target about camera position
campos	Set or get camera position
camproj	Set or get projection type
camroll	Rotate camera about viewing axis
camtarget	Set or get camera target
cameratoolbar	Control camera toolbar programmatically

camup	Set or get camera up-vector
camva	Set or get camera view angle
camzoom	Zoom camera in or out
view	Three-dimensional graph viewpoint specification
viewmtx	Generate view transformation matrices

Setting the Aspect Ratio and Axis Limits

daspect	Set or get data aspect ratio
pbaspect	Set or get plot box aspect ratio
xlim	Set or get the current x-axis limits
ylim	Set or get the current y-axis limits
zlim	Set or get the current z-axis limits

Object Manipulation

pan	Turns panning on or off
reset	Reset axis or figure
rotate	Rotate objects about specified origin and direction
rotate3d	Interactively rotate the view of a three-dimensional plot
selectmoveresize	Interactively select, move, or resize objects
zoom	Zoom in and out on a two-dimensional plot

Selecting Region of Interest

| dragrect | Drag XOR rectangles with mouse |
| rbbox | Rubberband box |

Lighting

camlight	Cerate or position Light
light	Light object creation function
lightangle	Position light in sphereical coordinates
lighting	Lighting mode
material	Material reflectance mode

Transparency

alpha	Set or query transparency properties for objects in current axes
alphamap	Specify the figure alphamap
alim	Set or query the axes alpha limits

Volume Visualization

coneplot	Plot velocity vectors as cones in three-dimensional vector field
contourslice	Draw contours in volume slice plane
curl	Compute curl and angular velocity of vector field
divergence	Compute divergence of vector field
flow	Generate scalar volume data
interpstreamspeed	Interpolate streamline vertices from vector-field magnitudes
isocaps	Compute isosurface end-cap geometry
isocolors	Compute colors of isosurface vertices
isonormals	Compute normals of isosurface vertices
isosurface	Extract isosurface data from volume data
reducepatch	Reduce number of patch faces

reducevolume	Reduce number of elements in volume data set
shrinkfaces	Reduce size of patch faces
slice	Draw slice planes in volume
smooth3	Smooth three-dimensional data
stream2	Compute two-dimensional stream line data
stream3	Compute three-dimensional stream line data
streamline	Draw stream lines from two- or three-dimensional vector data
streamparticles	Draws stream particles from vector volume data
streamribbon	Draws stream ribbons from vector volume data
streamslice	Draws well-spaced stream lines from vector volume data
streamtube	Draws stream tubes from vector volume data
surf2patch	Convert surface data to patch data
subvolume	Extract subset of volume data set
volumebounds	Return coordinate and color limits for volume (scalar and vector)

Creating Graphical User Interfaces Predefined Dialog Boxes

dialog	Create and display dialog box
errordlg	Create and display error dialog box
helpdlg	Create and display help dialog box
inputdlg	Create and display input dialog box
listdlg	Create and display list selection dialog box
msgbox	Create and display message dialog box
pagesetupdlg	Display page setup dialog box
printdlg	Display print dialog box
questdlg	Display question dialog box
uigetdir	Display standard dialog box for retrieving a directory
uigetfile	Display standard dialog box for retrieving files
uigetpref	Display dialog box for retrieving preferences
uiputfile	Display standard dialog box for saving files
uisave	Display standard dialog box for saving workspace variables
uisetcolor	Display standard dialog box for setting an object's ColorSpec
uisetfont	Display standard dialog box for setting an object's font characteristics
waitbar	Display waitbar
warndlg	Display warning dialog box

Deploying User Interfaces

guidata	Store or retrieve GUI data
guihandles	Create a structure of handles
movegui	Move GUI figure to specified location onscreen
openfig	Open new copy or raise existing copy of GUI figure

Developing User Interfaces

| guide | Start the GUI Layout Editor |
| inspect | Display Property Inspector |

Working with Application Data

| getappdata | Get value of application-defined data |

isappdata   True if application-defined data exists
rmappdata   Remove application-defined data
setappdata   Specify application-defined data
guidata   Store or retrieve GUI data

Interactive User Input
ginput   Graphical input from a mouse or cursor
waitfor   Wait for conditions before resuming execution
waitforbuttonpress   Wait for key/buttonpress over figure

Working with Preferences
addpref   Add preference
getpref   Get preference
ispref   Test for existence of preference
rmpref   Remove preference
setpref   Set preference
uigetpref   Display dialog box for retrieving preferences
uisetpref   Manage preferences used in **uigetpref**

User Interface Objects
menu   Generate menu of choices for user input
uibuttongroup   Create container object to exclusively manage radio buttons and toggle buttons
uicontextmenu   Create context menu
uicontrol   Create user interface control object
uimenu   Create menus on figure windows
uipanel   Create panel container object
uipushtool   Create push button on a toolbar
uitoggletool   Create toggle button on a toolbar
uitoolbar   Create toolbar on a figure

Finding Objects from Callbacks
findall   Find all graphics objects
findfigs   Find visible off-screen figures
findobj   Locate graphics objects with specific properties
gcbf   Return handle of figure containing object whose callback is executing
gcbo   Return handle of object whose callback is executing

GUI Utility Functions
selectmoveresize   Select, move, resize, or copy axes and uicontrol graphics objects
textwrap   Return wrapped string matrix for given uicontrol
uistack   Restack objects

Controlling Program Execution
uiresume   Resume program execution halted with **uiwait**
uiwait   Halt program execution, restart with **uiresume**

External Interfaces Dynamic Link Libraries
calllib   Call function in external library
libfunctions   Return information on functions in external library
libfunctionsview   Create window displaying information on functions in external library

libisloaded	Determine if external library is loaded
libpointer	Create pointer object for use with external libraries
libstruct	Construct structure as defined in external library
loadlibrary	Load external library into MATLAB
unloadlibrary	Unload external library from memory

Java

class	Create object or return class of object
fieldnames	Return property names of object
import	Add package or class to current Java import list
inspect	Display graphical interface to list and modify property values
isa	Determine if input is object of given class
isjava	Determine if input is Java object
ismethod	Determine if input is object method
isprop	Determine if input is object property
javaaddpath	Add entries to dynamic Java class path
javaArray	Construct Java array
javachk	Generate error message based on Java feature support
javaclasspath	Set and get dynamic Java class path
javaMethod	Invoke Java method
javaObject	Construct Java object
javarmpath	Remove entries from dynamic Java class path
methods	Display information on class methods
methodsview	Display information on class methods in separate window
usejava	Determine if Java feature is supported in MATLAB

Component Object Model and ActiveX

actxcontrol	Create ActiveX control in figure window
actxcontrollist	List all currently installed ActiveX controls
actxcontrolselect	Display graphical interface for creating ActiveX control
actxserver	Create COM Automation server
addproperty	Add custom property to object
class	Create object or return class of object
delete	Delete COM control or server
deleteproperty	Remove custom property from object
enableservice	Enable DDE or COM Automation server
eventlisteners	Return list of events attached to listeners
events	Return list of events the control can trigger
Execute	Execute MATLAB command in server
Feval	Evaluate MATLAB function in server
fieldnames	Return property names of object
get	Get property value from interface, or display properties
GetCharArray	Get character array from server
GetFullMatrix	Get matrix from server
GetVariable	Returns data from variable in server workspace
GetWorkspaceData	Get data from server workspace
inspect	Display graphical interface to list and modify property values

interfaces	List custom interfaces to COM server
invoke	Invoke method on object or interface, or display methods
isa	Detect object of given MATLAB class or Java class
iscom	Determine if input is COM object
isevent	Determine if input is event
isinterface	Determine if input is COM interface
ismethod	Determine if input is object method
isprop	Determine if input is object property
load	Initialize control object from file
MaximizeCommandWindow	Display server window on Windows desktop
methods	List all methods for control or server
methodsview	Display graphical interface to list method information
MinimizeCommandWindow	Minimize size of server window
move	Move or resize control in parent window
propedit	Display built-in property page for control
PutCharArray	Store character array in server
PutFullMatrix	Store matrix in server
PutWorkspaceData	Store data in server workspace
Quit	Terminate MATLAB server
registerevent	Register event handler with control's event
release	Release interface
save	Serialize control object to file
send	Obsolete—duplicate of events
set	Set object or interface property to specified value
unregisterallevents	Unregister all events for control
unregisterevent	Unregister event handler with control's event

Dynamic Data Exchange

ddeadv	Set up advisory link
ddeexec	Send string for execution
ddeinit	Initiate DDE conversation
ddepoke	Send data to application
ddereq	Request data from application
ddeterm	Terminate DDE conversation
ddeunadv	Release advisory link

Web Services

callSoapService	Send SOAP message off to endpoint
createClassFromWsdl	Create MATLAB object based on WSDL file
createSoapMessage	Create SOAP message to send to server
callSoapService	Convert response string from SOAP server to MATLAB type

Serial Port Devices

clear	Remove serial port object from MATLAB workspace
delete	Remove serial port object from memory
disp	Display serial port object summary information
fclose	Disconnect serial port object from the device

fgetl	Read from device and discard the terminator
fgets	Read from device and include the terminator
fopen	Connect serial port object to the device
fprintf	Write text to the device
fread	Read binary data from the device
fscanf	Read data from device and format as text
fwrite	Write binary data to the device
get	Return serial port object properties
instrcallback	Display event information when an event occurs
instrfind	Return serial port objects from memory to the MATLAB workspace
isvalid	Determine if serial port objects are valid
length	Length of serial port object array
load	Load serial port objects and variables into MATLAB workspace
readasync	Read data asynchronously from the device
record	Record data and event information to a file
save	Save serial port objects and variables to MAT-file
serial	Create a serial port object
serialbreak	Send break to device connected to the serial port
set	Configure or display serial port object properties
size	Size of serial port object array
stopasync	Stop asynchronous read and write operations

C Programs

MEX-Files	Perform operations in the MATLAB environment f rom your C MEX-files
MATLAB Engine	Call MATLAB from your own C programs
MX Array Manipulation	Create and manipulate MATLAB arrays from C MEX and Engine routines
MAT-File Access	Incorporate and use MATLAB data in your own C programs

FORTRAN Programs

MEX-Files	Perform operations in the MATLAB environment from your FORTRAN MEX-files
MATLAB Engine	Call MATLAB from your own FORTRAN programs
MX Array Manipulation	Create and manipulate MATLAB arrays from FORTRAN MEX and Engine routines
MAT-File Access	Incorporate and use MATLAB data in your own FORTRAN programs

Bibliography

Chapman, S. J., *MATLAB Programming for Engineers*, 3rd ed., Thomson Engineering, 2005.

Etter, D., Kuncicky, D., and Moore, H., *Introduction to MATLAB 7*, Prentice-Hall, Upper Saddle River, NJ, 2005.

Gilat, A., *MATLAB: An Introduction with Applications*, 2nd ed., Wiley, New York, 2005.

Hanselman, D. C., and Littlefield, B. L., *Mastering MATLAB 7*, Prentice-Hall, Upper Saddle River, NJ, 2005.

Magrab, E. B., Azarm, S., Balachandran, B., Duncan, J., Herold, K., and Walsh, G., *An Engineer's Guide to MATLAB*, with Applications from Mechanical, Aerospace, Electrical, and Civil Engineering, 2nd ed., Prentice-Hall, Upper Saddle River, NJ, 2005.

Palm, W. J., III, *Introduction to MATLAB 7 for Engineers*, McGraw-Hill, New York, 2005.

Stanley, W. D., *Technical Analysis and Applications with MATLAB*, Thomson Delmar Learning, 2005.

Aerospace Blockset User's Guide, The MathWorks, Natick, MA, 3 Aug. 2006.

Bioinformatics Toolbox User's Guide, The MathWorks, Natick, MA, 3 Aug. 2006.

Carlson, E. S., *Efficient MATLAB for Engineers*, OtFringe, 2004.

CDMA Reference Blockset User's Guide, The MathWorks, Natick, MA, 3 Aug. 2006.

Communications Blockset User's Guide, The MathWorks, Natick, MA, 3 Aug. 2006.

Communications Toolbox User's Guide, The MathWorks, Natick, MA, 3 Aug. 2006.

Control System Toolbox User's Guide, The MathWorks, Natick, MA, 3 Aug. 2006.

Curve Fitting Toolbox User's Guide, The MathWorks, Natick, MA, 3 Aug. 2006.

Dabney, J. B., and Harman, T. L., *Mastering Simulink*, Prentice-Hall, Upper Saddle River, NJ, 2004.

Data Acquisition Toolbox User's Guide, The MathWorks, Natick, MA, 3 Aug. 2006.

Database Toolbox User's Guide, The MathWorks, Natick, MA, 3 Aug. 2006.

Datafeed Toolbox User's Guide, The MathWorks, Natick, MA, 3 Aug. 2006.

Distributed Computing Toolbox User's Guide, The MathWorks, Natick, MA, 3 Aug. 2006.

Embedded Target for Infineon C166® Microcontrollers User's Guide, The MathWorks, Natick, MA, 3 Aug. 2006.

Embedded Target for Motorola® HC12 User's Guide, The MathWorks, Natick, MA, 3 Aug. 2006.

Embedded Target for Motorola® MPC555 User's Guide, The MathWorks, Natick, MA, 3 Aug. 2006.

473

Embedded Target for OSEK/VDX® User's Guide, The MathWorks, Natick, MA, 3 Aug. 2006.

Embedded Target for TI C2000™ DSP User's Guide, The MathWorks, Natick, MA, 3 Aug. 2006.

Embedded Target for TI C6000™ DSP User's Guide, The MathWorks, Natick, MA, 3 Aug. 2006.

Excel Link User's Guide, The MathWorks, Natick, MA, 3 Aug. 2006.

Extended Symbolic Math Toolbox User's Guide, The MathWorks, Natick, MA, 3 Aug. 2006.

Filter Design HDL Coder User's Guide, The MathWorks, Natick, MA, 3 Aug. 2006.

Filter Design Toolbox User's Guide, The MathWorks, Natick, MA, 3 Aug. 2006.

Financial Derivatives Toolbox User's Guide, The MathWorks, Natick, MA, 3 Aug. 2006.

Financial Time Series Toolbox User's Guide, The MathWorks, Natick, MA, 3 Aug. 2006.

Financial Toolbox User's Guide, The MathWorks, Natick, MA, 3 Aug. 2006.

Fixed-Income Toolbox User's Guide, The MathWorks, Natick, MA, 3 Aug. 2006.

Fixed-Point Toolbox User's Guide, The MathWorks, Natick, MA, 3 Aug. 2006.

Fuzzy Logic Toolbox User's Guide, The MathWorks, Natick, MA, 3 Aug. 2006.

GARCH Toolbox User's Guide, The MathWorks, Natick, MA, 3 Aug. 2006.

Gauges Blockset User's Guide, The MathWorks, Natick, MA, 3 Aug. 2006.

Genetic Algorithm and Direct Search Toolbox User's Guide, The MathWorks, Natick, MA, 3 Aug. 2006.

Image Acquisition Toolbox User's Guide, The MathWorks, Natick, MA, 3 Aug. 2006.

Image Processing Toolbox User's Guide, The MathWorks, Natick, MA, 3 Aug. 2006.

Instrument Control Toolbox User's Guide, The MathWorks, Natick, MA, 3 Aug. 2006.

Link for Code Composer Studio™ User's Guide, The MathWorks, Natick, MA, 3 Aug. 2006.

Link for ModelSim® User's Guide, The MathWorks, Natick, MA, 3 Aug. 2006.

Mapping Toolbox User's Guide, The MathWorks, Natick, MA, 3 Aug. 2006.

MATLAB® Builder for COM User's Guide, The MathWorks, Natick, MA, 3 Aug. 2006.

MATLAB® Builder for Excel User's Guide, The MathWorks, Natick, MA, 3 Aug. 2006.

MATLAB® Compiler User's Guide, The MathWorks, Natick, MA, 3 Aug. 2006.

MATLAB® Distributed Computing Engine User's Guide, The MathWorks, Natick, MA, 3 Aug. 2006.

MATLAB® Report Generator User's Guide, The MathWorks, Natick, MA, 3 Aug. 2006.

MATLAB® User's Guide, The MathWorks, Natick, MA, 3 Aug. 2006.

MATLAB® Web Server User's Guide, The MathWorks, Natick, MA, 3 Aug. 2006.

Model-Based Calibration Toolbox User's Guide, The MathWorks, Natick, MA, 3 Aug. 2006.

Model Predictive Control Toolbox User's Guide, The MathWorks, Natick, MA, 3 Aug. 2006.

Neural Network Toolbox User's Guide, The MathWorks, Natick, MA, 3 Aug. 2006.

OPC Toolbox User's Guide, The MathWorks, Natick, MA, 3 Aug. 2006.

Optimization Toolbox User's Guide, The MathWorks, Natick, MA, 3 Aug. 2006.

Partial Differential Equation Toolbox User's Guide, The MathWorks, Natick, MA, 3 Aug. 2006.

Real-Time Windows Target User's Guide, The MathWorks, Natick, MA, 3 Aug. 2006.

Real-Time Workshop® Embedded Coder User's Guide, The MathWorks, Natick, MA, 3 Aug. 2006.

Real-Time Workshop® User's Guide, The MathWorks, Natick, MA, 3 Aug. 2006.

RF Blockset User's Guide, The MathWorks, Natick, MA, 3 Aug. 2006.

RF Toolbox User's Guide, The MathWorks, Natick, MA, 3 Aug. 2006.

Robust Control Toolbox User's Guide, The MathWorks, Natick, MA, 3 Aug. 2006.

Signal Processing Blockset User's Guide, The MathWorks, Natick, MA, 3 Aug. 2006.

Signal Processing Toolbox User's Guide, The MathWorks, Natick, MA, 3 Aug. 2006.

SimDriveline User's Guide, The MathWorks, Natick, MA, 3 Aug. 2006.

SimMechanics User's Guide, The MathWorks, Natick, MA, 3 Aug. 2006.

SimPowerSystems User's Guide, The MathWorks, Natick, MA, 3 Aug. 2006.

Simulink® Accelerator User's Guide, The MathWorks, Natick, MA, 3 Aug. 2006.

Simulink® Control Design User's Guide, The MathWorks, Natick, MA, 3 Aug. 2006.

Simulink® Fixed Point User's Guide, The MathWorks, Natick, MA, 3 Aug. 2006.

Simulink® Parameter Estimation User's Guide, The MathWorks, Natick, MA, 3 Aug. 2006.

Simulink® Report Generator User's Guide, The MathWorks, Natick, MA, 3 Aug. 2006.

Simulink® Response Optimization User's Guide, The MathWorks, Natick, MA, 3 Aug. 2006.

Simulink® User's Guide, The MathWorks, Natick, MA, 3 Aug. 2006.

Simulink® Verification and Validation User's Guide, The MathWorks, Natick, MA, 3 Aug. 2006.

Spline Toolbox User's Guide, The MathWorks, Natick, MA, 3 Aug. 2006.

Stateflow Coder User's Guide, The MathWorks, Natick, MA, 3 Aug. 2006.

Stateflow® User's Guide, The MathWorks, Natick, MA, 3 Aug. 2006.

Statistics Toolbox User's Guide, The MathWorks, Natick, MA, 3 Aug. 2006.

Symbolic Math Toolbox User's Guide, The MathWorks, Natick, MA, 3 Aug. 2006.

System Identification Toolbox User's Guide, The MathWorks, Natick, MA, 3 Aug. 2006.

Video and Image Processing Blockset User's Guide, The MathWorks, Natick, MA, 3 Aug. 2006.

Virtual Reality Toolbox User's Guide, The MathWorks, Natick, MA, 3 Aug. 2006.

Wavelet Toolbox User's Guide, The MathWorks, Natick, MA, 3 Aug. 2006.

xPC TargetBox® User's Guide, The MathWorks, Natick, MA, 3 Aug. 2006.

xPC TargetBox® Embedded Option User's Guide, The MathWorks, Natick, MA, 3 Aug. 2006.

xPC TargetBox® User's Guide, The MathWorks, Natick, MA, 3 Aug. 2006.

Lyshevski, S. E., *Engineering and Scientific Computations Using MATLAB*, Wiley, New York, 2003.

Hunt, B. R., Lipsman, R. L., Rosenberg, J. M., Coombes, K. R., Osborn, J. E., and Stuck, G. J., *A Guide to MATLAB: For Beginners and Experienced Users*, Cambridge Univ. Press, 2001.

Palm, W. J., III, *MATLAB for Engineering Applications*, WCB/McGraw-Hill, New York, 1999.

Etter, D. M., *Engineering Problem Solving with MATLAB*, 2nd ed., Prentice-Hall, Upper Saddle River, NJ, 1997.

Etter, D. M., and Kuncicky, D., *Introduction to MATLAB for Engineers and Scientists*, Prentice-Hall, Upper Saddle River, NJ, 1996.

Index

Supporting Materials

To download your software and any other software updates, please go to http://www.aiaa.org/publications/supportmaterials. Follow the instructions provided and enter the following password: **mfiles.** For a complete listing of titles in the AIAA Education Series, as well as other AIAA publications, please visit **http://www.aiaa.org.**